LECTURE NOTES ON

THE MATHEMATICAL THEORY OF GENERALIZED BOLTZMANN MODELS

Series on Advances in Mathematics for Applied Sciences – Vol. 51

LECTURE NOTES ON
THE MATHEMATICAL THEORY OF GENERALIZED BOLTZMANN MODELS

Nicola Bellomo
Dipartimento di Matematica, Politecnico di Torino, Italy

Mauro Lo Schiavo
Dipartimento Metodi e Modelli per le Applicazioni
Università Roma, Italy

World Scientific
Singapore • New Jersey • London • Hong Kong

Published by

World Scientific Publishing Co. Pte. Ltd.
5 Toh Tuck Link, Singapore 596224
USA office: 27 Warren Street, Suite 401-402, Hackensack, NJ 07601
UK office: 57 Shelton Street, Covent Garden, London WC2H 9HE

Library of Congress Cataloging-in-Publication Data
Bellomo, N.
Lecture notes on the mathematical theory of generalized Boltzmann models / Nicola Bellomo, Mauro Lo Schiavo.
p. cm. — (Series on advances in mathematics for applied sciences ; vol. 51)
Includes bibliographical references.
ISBN-13 978-981-02-4078-3
ISBN-10 981-02-4078-3
1. Maxwell-Boltzmann distribution law. 2. Kinetic theory of matter -- Mathematical models. I. Schiavo, Mauro Lo. II. Title.
III. Title: Mathematical theory of generalized Boltzmann models.
IV. Series.
QC175.16.B6B45 1999
530.13'6--dc21 99-40190
CIP

British Library Cataloguing-in-Publication Data
A catalogue record for this book is available from the British Library.

Contents

Preface . . . . **xi**

Chapter 1. Generalized Kinetic Models **1**

1.1 Introduction 1

1.2 Generalized Kinetic Models 3

1.3 Generalized Models and Plan of the Book 11

1.4 Aim of the Book 16

1.5 References 17

Chapter 2. Mathematical Background: Measure, Integration, Topology **21**

2.1 Introduction 21

2.2 Tools from Measure Theory 22

2.2.1 Physical connections 23

2.2.2 Basics on measure theory 25

2.2.3 Measures on the real line 30

2.2.4 Measures on $\mathbb{R}^n$ 35

2.3 Tools from Integration Theory 40

2.3.1 Lebesgue integrals 40

2.3.2 Main theorems of integration theory 42

2.3.3 Distribution functionals 49

2.3.4 Some notable generalizations . . 56

2.3.5 Two conclusive examples 61

2.4 Tools from Topology 68

2.4.1 Basics on topologies 68

2.4.2 Function spaces 74

2.4.3 The linear spaces case 79

2.5 References 84

Chapter 3. Models of Population Dynamics with Stochastic Interactions . . . 87

3.1 Introduction 87

3.2 The Generalized Jager and Segel Model . 89

3.3 On the Initial Value Problem 93

3.4 Stationary Points 97

3.5 Applications and Perspectives 100

3.5.1 Modelling η and ψ 100

3.5.2 A model of social behaviors . . . 103

3.5.3 A model in epidemiology 106

3.5.4 Perspectives 108

3.6 References 110

Chapter 4. Generalized Kinetic Models for Coagulation and Fragmentation . . 113

4.1 Introduction 113

4.2 Description of the Models 116

4.3 Mathematical Problems 120

4.3.1 Existence of solutions 121

4.3.2 Equilibrium solutions and stability 126

4.4 Critical Analysis and Perspectives . . . 131

4.5 References 134

Chapter 5. Kinetic Cellular Models in the Immune System Competition 137

5.1 Kinetic Models Towards Immunology . . 137

5.2 Scaling in Kinetic Cellular Models . . . 140

5.3 Phenomenological System and Modelling 144

5.4 Kinetic Evolution Equations 149

5.5 Qualitative Analysis, Applications, and Perspectives 158

5.5.1 Qualitative analysis 159

5.5.2 Simulation problems 161

5.5.3 Perspectives 165

5.6 References 167

Chapter 6. Kinetic Models for the Evolution of Antigens Generalized Shape 171

6.1 An Introduction to the Generalized Shape 171

6.2 The Mathematical Model 173

6.3 On the Initial Value Problem 177

6.4 Applications and Developments 183

6.5 References 186

Chapter 7. The Boltzmann Model 189

7.1 Introduction 189

7.2 The Nonlinear Boltzmann Equation . . 195

7.3 Mathematical Problems 199

7.4 Analytic Treatment 205

7.4.1 The Cauchy problem for small initial data 206

7.4.2 The Cauchy problem for large initial data 211

7.4.3 The initial-boundary value problem 216

7.4.4 Open problems 216

7.4.5 Evolution problems in the presence of a force field 218

7.4.6 Shock waves 219

7.4.7 Asymptotic analysis 219

7.5 Computational Methods 220

7.6 References 224

Chapter 8. Generalized Kinetic Models for Traffic Flow 235

8.1 Introduction 235

8.2 Traffic Flow and Hydrodynamics 236

8.2.1 Scalar hydrodynamic models . . 240

8.2.2 Vector hydrodynamic models . . 243

8.3 Kinetic Traffic Flow Models 246

8.3.1 From Prigogine's to Paveri Fontana's modelling . . . 249

8.3.2. Developments in kinetic modelling 254

8.3.3 Evolution problems 262

8.4 Perspectives 265

8.5 References 268

Chapter 9. Dissipative Kinetic Models for Disparate Mixtures 273

9.1 Introduction 273

9.2 Dissipative Collision Dynamics 275

9.2.1 Cluster conservative collisions . . 276

9.2.2 Cluster destructive collisions . . 278

9.3 Kinetic Equations for Mixtures of Clusters 282

9.4 Mixtures with Continuous Mass Distribution 286

9.5 Mathematical Problems 288

9.6 Perspectives in Modelling 289
9.7 References 290

Chapter 10. Research Perspectives 293

10.1 Introduction 293
10.2 Discrete Generalized Models 295
10.2.1 The discrete Boltzmann equation 297
10.2.2 Discrete models in immunology . 302
10.3 Looking for a General Structure 315
10.4 Development of New Models 323
10.5 Closure 325
10.6 References 326

Books and Review Papers 329

Preface

Mathematical models in applied sciences are generally obtained by developments of the traditional models of phenomenologic continuum mechanics or, more in general, of mathematical physics. A large part of these models are stated in terms of partial differential equations. Solutions and simulations are related to initial and/or boundary value problems. Nonlinearity is a general feature of models, linearity is a special case often related to an artificial simplification of physical reality.

The literature on nonlinear partial differential equations in mathematical physics and applied sciences is vast and specialized over different aspects, e.g. modelling, qualitative analysis, numerical and computational methods. On the other hand, the modelling of physical systems in nonequilibrium statistical mechanics is generally based on Boltzmann-like models. The mathematical model is an evolution equation for the dependent variable that, in this case, is the distribution of a large ensemble of interacting particles.

In particular, the Boltzmann model is an evolution equation for one-particle distribution function of a diluted gas of identical particles, which are modelled by point masses identified relatively to space and velocity variables. Macroscopic observable quantities are recovered by suitable averages with respect to the distribution function.

This book is based on the idea that Boltzmann-like modelling methods can be developed to design, with special attention to applied sciences, kinetic-type models which will be called ***generalized kinetic models***. In particular, these models consist in evolution equations for the statistical distribution over the physical state of each individual of a large population. The evolution is determined both by interactions among individuals or by external actions.

The mathematical structure of this class of models is similar to that of the Boltzmann equation. Statistical measures of the distribution function may define the evolution of observable quantities.

Considering that generalized kinetic models can play an important role in dealing with several interesting systems in applied sciences, this book provides a unified presentation of this topic with direct reference to modelling, mathematical statement of problems, qualitative and computational analysis, and applications.

Models reported and proposed in this book refer to several fields of natural, applied and technological sciences. Population dynamics and socio-economical behaviors are dealt with in Chapter 3. Models of aggregation and fragmentation phenomena are dealt with in Chapter 4. Models of biology and immunology are considered in Chapters 5 and 6. The Boltzmann equation is dealt with in Chapter 7 as a particular, may be the fundamental, reference model of the large variety of kinetic models reported in this book. Traffic flow models are dealt with in Chapter 8. Various aspects on modelling mixture of particles undergoing classic and dissipative interactions are proposed in Chapter 9. The last chapter is devoted to a survey of open problems and research perspectives. A collective bibliography of books and review papers is reported at the end of the book.

A more detailed index can be recovered in the last section of the first chapter motivated by a preliminary description of the main features of what are here called generalized kinetic models.

Although this book collects several research contributions dispersed in a vast literature, completeness is not claimed. The contents are developed with methodological aims, with an effort to cover, at least, the following topics:

i) Modelling procedures related to specific fields of applied sciences.

ii) Qualitative analysis of mathematical problems.

iii) A survey of the applications already available in the literature followed by some analysis of research perspectives in the field.

The contents have been motivated also by some mathematicians who have directly or indirectly stimulated this effort. In particular, we wish to thank Lee Segel, who presented in one of the Oberwolfach meetings, in 1990, an interesting model (proposed in a paper with Eva Jager) on population dynamics with kinetic interaction. This model and some stimulating discussions gave the first hint to deal with the content of this book. The above model has been chosen here as the main example of the link between Boltzmann models and generalized kinetic models. Further motivations were given in a workshop on generalizations of the Boltzmann equation organized by Helmut Neunzert in Kaiserslautern in 1991. There, the relevance of developing kinetic models in the applied science was clearly posed. Finally, the first author is indebted to his colleague and friend Antonio Romano, who organized a Ph.D. course on the topics dealt with in this book. The Lectures given at the University of Napoli can be regarded as the preliminary background of this book.

The authors acknowledge the support of the National Research Council, C.N.R., under contract 98.03633.st74.

Nicola Bellomo and Mauro Lo Schiavo

Chapter 1

Generalized Kinetic Models

1.1 Introduction

This book deals with the derivation, analysis and applications of generalized kinetic models, and it describes how they may be referred in various fields of applied sciences, say astrophysics, traffic flow, mathematical biology, theoretical immunology, social sciences, etc.

The literature on these models, that are characterized by a mathematical structure somewhat similar to the nonlinear Boltzmann equation, is generally dispersed in several specialized areas. On the contrary, we are concerned with a unified presentation of the whole, large and highly general, class of models among which the Boltzmann equation can be regarded as a particular, however relevant, example.

As it is well known [CEa], [TRa], the Boltzmann equation is an evolution equation for the one-particle distribution function of a diluted gas of identical particles, modelled by point masses and identified relatively to space and velocity variables. Macroscopic observable quantities can be recovered by convenient moments weighted by the distribution function. Similarly, a ***generalized Boltzmann model*** consists in an evolution equation for the (one individual) statistical ***distribution function*** over the ***physical state*** of individuals in a large population. The distribution function, and the related moments, are supposed to describe and characterize the system under investigation.

Indeed, generalized models are derived by means of arguments similar to those used in the heuristic (phenomenological) derivation of the Boltzmann equation. They consist in an evolution equation based on a balancing related, in general, to pairwise interactions between individuals of the populations.

In this framework, the ***microscopic description*** refers to single individuals and to their interactions, whereas the ***macroscopic description*** refers to overall behaviors and to observable quantities averaged over great numbers of individuals. Macroscopic observable quantities may be obtained, in the mathematical model, as moments with respect to the distribution function.

In the Boltzmann equation, the distribution function is interpreted as the number density of particles to be found in a selected element of the state space at a given time. Similarly, the distribution function in traffic flow models refers to the number of cars that, at a certain time, are likely to be found in an elementary tract of a road. Analogously, in (spatially homogeneous) cellular dynamics, the distribution function represents the number of cells per unit volume that are expected at a certain time to exhibit a certain value of the cellular activation.

As for the Boltzmann equation, in the above models one may suppose that the size of the cars or of the cells is negligible with respect to the characteristic dimensions of the system: say the length of the road in the case of the traffic flow model, or the overall dimensions of the vertebrate in the biological system. Still, this assumption is not always applied or necessary. Indeed, it is well known that the Enskog equation is a model, similar to the Boltzmann equation, with the assumption that the dimension of the particles is finite, see [BLa].

The contents of this book is developed bearing in mind that generalized kinetic models may be useful tools in describing a large class of complex physical systems. Moreover, mathematical problems related to the analysis of these models are often an interesting and challenging task for applied mathematicians.

This first chapter is an introduction to the subject of the book, and includes a presentation of its contents. In more detail, this chapter is divided into three sections:

Section 1.1 is the introduction.

Section 1.2 provides a definition of generalized models of the Boltzmann type, or ***generalized kinetic models***. It also describes some of the lines that should be followed for their derivation.

Section 1.3 contains a concise guide to the contents of the book and to the arguments that will be developed in the chapters that follow. Each chapter deals with a certain class of models and shows how it is related to various fields of applied sciences.

1.2 Generalized Kinetic Models

Here we describe the characteristic features that generalized kinetic models have in common. Moreover, we indicate some of the rules that may be convenient to follow in their constructions. The hints that will be given have to be interpreted as preliminary to the detailed analysis that will be developed in the chapters that follow. Obviously, each particular class of models will differ slightly from the general framework that is proposed here.

A model will be called, according to [BLc], a ***generalized Boltzmann model*** or, alternatively, a ***generalized kinetic model*** when it is characterized by the following features:

- The system is constituted by a large number of ***interacting individuals***, or ***objects***. For mass conservative systems, the total number of individuals is a constant; for nonconservative systems it varies with time.
- The physical state of each individual is labeled by a state variable $\mathbf{u}$, which in general is a vector defined in some domain $\mathcal{D}_{\mathbf{u}}$ of $\mathbf{R}^n$. As such, state variable $\mathbf{u}$ may, or may not, include the location $\mathbf{x}$ and/or the velocity $\mathbf{v}$.

- The overall state of the system is described by a (statistical) distribution function

$$f = f(t, \mathbf{u})\,; \quad t \in [0, T]\,, \quad \mathbf{u} \in \mathcal{D}_{\mathbf{u}} \subseteq \mathbf{R}^n\,, \quad f(t, \mathbf{u}) \geq 0\,, \tag{1.2.1}$$

that assigns, by means of the expectation value:

$$N(t\,;\, \mathbf{u} \in I_{\mathbf{u}}) = \int_{I_{\mathbf{u}}} f(t, \mathbf{u})\, d\mathbf{u}\,, \tag{1.2.2}$$

the number of individuals that at time t is expected to be found in a state lying in the set $I_{\mathbf{u}} \subseteq \mathcal{D}_{\mathbf{u}}$. When the total number of individuals is *a priori* finite, for instance when it is constant with respect to time, then the distribution function f may acquire the structure of a probability density, and its integral over the whole range $\mathcal{D}_{\mathbf{u}} \subseteq \mathbf{R}^n$ of $\mathbf{u}$ normalized to one:

$$\int_{\mathcal{D}_{\mathbf{u}}} f(t, \mathbf{u})\, d\mathbf{u} = 1\,, \qquad \forall t \in [0, T]\,. \tag{1.2.3}$$

- The physical state of each individual is modified by pairwise interactions. Interaction rates may depend on the states of the interacting individuals, and can generate not only a change in the populations of the various states, yet preserving the population total number, but also proliferation or destruction of individuals.
- The dynamics of the system is ruled by an ***evolution equation*** on the distribution function f. It is obtained by equating the time derivative of f to the difference between a ***gain*** and a ***loss*** term. The gain term takes into account the individuals that reach the state $\mathbf{u}$ because of interactions, or of some natural or external mechanisms driving the individual towards the state $\mathbf{u}$. The loss term is related to the individuals that, because of similar reasons, leave the state $\mathbf{u}$.

- Interactions between pair of individuals are not correlated, and hence the probability densities (and, for large N, the distributions) satisfy

$$f_2(t, \mathbf{u}_1, \mathbf{u}_2) = f(t, \mathbf{u}_1) f(t, \mathbf{u}_2) . \tag{1.2.4}$$

As in the case of the Boltzmann equation, the last assumption, which is required in the derivation of almost all the models, can only be proved for very special conditions; again it has to be regarded as an approximation of physical reality.

Well posed mathematical problems, generally initial and/or boundary value problems, are the problems whose (unique) solution provides the explicit time dependence of the probability density f. Subsequently the ***moments***, i.e., the averages of convenient functions of the microscopic variables weighted by f, provide a (macroscopic statistical) information on the state of the system, and in some cases can be related to macroscopic quantities. This calculation may require the background of sophisticated asymptotic analysis [LAa].

For some generalized models, the analytic discussion and computational problems can be far less complicated than the analogous ones related to the pure Boltzmann model. This happens when the evolution problem for generalized systems preserves the L_1-norm of f, a property that naturally allows global existence proofs and computational shortcuts towards quantitative information on the behavior of the system.

Unfortunately, kinetic modelling refers to extremely different physical situations, and may have very weak resembling with the kinetic theory of gases. Therefore, the model can be completely defined only in few specific cases. This is why additional details about the common lines that characterize generalized kinetic models cannot be given at this stage, and will be provided only in the following chapters.

In particular, the assumptions summarized above to provide an overall description of the generalized kinetic models are based on microscopic interactions, and these in general may modify the evolution of the distribution function. This is especially true for those models that are more closely related to kinetic theory. On some other instances, the evolution equation may be heuristically derived without a detailed description of the said interactions. Then we shall talk about ***phenomenologic generalized kinetic models.*** In some other cases, generalized kinetic models are derived using both the microscopic and the phenomenologic modelling. This is what happens, as we shall see, for traffic flow models.

An example will now be given to illustrate at least a class of generalized kinetic models, and be able to examine its mathematical structure and relationships with other classes of models.

Consider the kinetic model proposed by Jager and Segel [JAa] to describe the social behavior in a large population of anonymous interacting insects. The physical motivations that generate the model are not discussed here (this matter will be dealt with in Chapter 3). Instead, our attention will be focused on the mathematical structure of the model. In particular, we recall that the model is defined by the following assumptions:

Assumption 1.2.1. *Each individual is characterized by a social state called* ***dominance****, denoted by a variable u which takes values in the interval* $[0, 1]$.

Assumption 1.2.2. *No individual with state v changes his state until an encounter with another individual occurs.*

Assumption 1.2.3. *A function $\eta(v, w)$ can be properly defined to account for the rate of encounters between individuals with states v and w, respectively.*

Assumption 1.2.4. *A probability density $\psi(v, w; u)$ can be identified to account for the probability that after an encounter between an individual with state v and one with state w, the subject in state v transits to a state in an interval containing u.*

The mathematical model defines the evolution of the one-particle probability density function $f = f(t, u)$ relative to the state u. The model is obtained by equating the time derivative of f to the difference between the ***gain*** of individuals that in the unit time reach state u and the ***loss*** of individuals that abandon state u. These assumptions, linked to further hypotheses of statistical independence of the multiple distributions, and of time independence of the total number of individuals, yield the following model:

$$\begin{aligned}\frac{\partial f}{\partial t}(t,u) &= J[f](t,u) = G[f](t,u) - L[f](t,u)\\ &= \int_0^1 \int_0^1 \eta(v,w)\psi(v,w;u) f(t,v) f(t,w)\,dv\,dw\\ &\quad - f(t,u)\int_0^1 \eta(u,w) f(t,w)\,dw\,. \end{aligned} \tag{1.2.5}$$

Though being an important reference point, this model is a general one and does not include many other possibilities. For instance, in some cases interactions may not be described in the framework of Assumption 1.2.4. Or, the system may undergo multiple interactions, and so on. On the other hand, the largest part of the models dealt with in this book is very similar to the above one. Therefore, the model by Jager and Segel will be an important reference for the future contents.

Technical generalizations are immediate. For instance one can deal with models wherein the state variable is a vector $\mathbf{u}$. In this case, the model becomes

$$f = f(t,\mathbf{u}) : \qquad \frac{\partial f}{\partial t} = J[f]\,, \tag{1.2.6}$$

where $\mathbf{u}$ defines the multiple valued state variable of the individuals in the system. Space dependent generalized models are those where a space variable $\mathbf{x}$ must be included in the variable $\mathbf{u}$.

Consider at first the spatially homogeneous case. A mathematical modelling, totally similar to the Jager and Segel model, can be developed when the following two quantities are recovered:

- The ***encounter rate*** $\eta(\mathbf{v}, \mathbf{w})$, defining the number of encounters per unit time in the unit volume between subjects with states $\mathbf{v}$ and $\mathbf{w}$.
- The ***transition probability density*** $\psi(\mathbf{v}, \mathbf{w}; \mathbf{u})$, denoting the probability density that a subject with state $\mathbf{v}$ ends up in state $\mathbf{u}$ because of an encounter with a subject in state $\mathbf{w}$.

These two quantities are, in general, deduced by an analysis that is typical of the system under consideration. In the case of the Boltzmann equation, for example, the functions η and ψ are obtained in the framework of classical mechanics. In particular, when $\mathbf{u}$ contains the velocity of the particle, η may be linked to the relative velocity of two encountering particles, whereas ψ is related to the mass, momentum, and energy conservation equations. Still further generalizations are immediately obtained. For instance, one may make use of mechanics in a broad sense, e.g., stochastic or quantum mechanics.

On the other hand, for some generalized models such as those about social or economical dynamics, the rate η may be constructed only on the basis of rules that are proper of the model. In the cited cases, they consist in laws concerning social and/or economic affinity within the same or among different populations. Furthermore, the probability ψ may be determined only by the use of models of micro-economy.

In all cases, independent of how the above terms are obtained, the evolution equation is given, as in the Jager and Segel case, by the balance

$$f = f(t, \mathbf{u}) : \quad \frac{\partial f}{\partial t} = G[f] - L[f], \tag{1.2.7}$$

where the balance is characterized by the ***gain*** term G, namely the number of individuals that reach state $\mathbf{u}$ after an encounter with other individuals, minus the ***loss*** term L, namely the number of individuals that abandon the state $\mathbf{u}$ after an encounter with other individuals.

Only binary encounters are considered if one accepts (as usual in kinetic theory) the assumption of statistical independence:

$$f_2(t,\mathbf{v},\mathbf{w}) = f(t,\mathbf{v})f(t,\mathbf{w})\,. \tag{1.2.8}$$

Hence the gain and loss terms assume the following expressions

$$G[f](t,\mathbf{u}) = \int_{\mathcal{D}_\mathbf{u}\times\mathcal{D}_\mathbf{u}} \eta(\mathbf{v},\mathbf{w})\psi(\mathbf{v},\mathbf{w};\mathbf{u})f(t,\mathbf{v})f(t,\mathbf{w})\,d\mathbf{v}\,d\mathbf{w}\,, \tag{1.2.9}$$

where $\mathcal{D}_\mathbf{u}$ is the domain of definition of $\mathbf{u}$, and

$$L[f](t,\mathbf{u}) = f(t,\mathbf{u})\int_{\mathcal{D}_\mathbf{u}} \eta(\mathbf{u},\mathbf{v})f(t,\mathbf{v})\,d\mathbf{v}\,. \tag{1.2.10}$$

These are the gain and loss terms that are to be substituted into Eq. (1.2.7) to obtain the evolution equation which, in fact, does not substantially differs from model (1.2.5).

As a further step, consider now the more general, spatially dependent case. According to it, one of the arguments of the distribution function f is the space variable $\mathbf{x}$. Although the structure of the evolution equation is still that indicated in Eq. (1.2.7), now the term on the left-hand side, that is the time derivative of f, contains as in the Boltzmann equation the flux of individuals related to the spatial gradient of f. As well, the term on the right-hand side, that still consists of a balance between the gain and loss terms, now depends on $\mathbf{x}$, and even the interactions may be affected by the space variables.

In all cases, the mathematical problems formulation follows the equation structure. The space independent models evolution is characterized by ***initial value problems*** obtained by linking the evolution equation to suitable initial conditions $f(t=0,\mathbf{u})$. The space dependent models evolution is ruled by ***initial-boundary value problems*** obtained by linking the evolution equation to suitable initial and boundary conditions. As in the case of Boltzmann equation, boundary conditions for generalized kinetic models may be defined by taking into account the physics of the system. For instance, flux or reflection conditions may be set at the boundaries of the space variable domain.

This book reports a review on the mathematical results about the initial and initial-boundary value problems for generalized kinetic models. Often, the techniques developed for such an analysis are very similar to those needed for the Boltzmann equation, see [BLb] and [GLa]. Instead, as we shall see, in other cases the analysis requires much simpler approaches. For some models, additional difficulties have to be tackled and new methods developed. In all cases, the mathematical analysis is developed keeping in mind that, to implement reasonable simulations, numerical and computational methods are ultimately needed. In other words, actual applicability of the model to real situations and problems of the applied sciences is an important feature of this book.

As already mentioned, some specific kinetic models may indeed be developed with a slightly different structure from that one reported in Eq. (1.2.7). For instance, source terms or multiple interactions may be included; this will be done in some of the models developed later on. However, the general structure of the evolution equation will be the same as that of Eq. (1.2.7), the main differences being those directly related to the particular state variable $\mathbf{u}$ and to the interactions among individuals.

Finally, we remark that all the models that we have mentioned so far are based on a derivation that starts from microscopic interactions. In some cases, however, one can design the model only if use is made of heuristic phenomenologic arguments. For instance, if the system is characterized by an equilibrium configuration f_e and shows a trend towards equilibrium, one can model the system as follows

$$f = f(t, \mathbf{u}) : \qquad \frac{\partial f}{\partial t} = \nu(f - f_e) \,, \tag{1.2.11}$$

that is, with an extremely simple linear assumption provided that ν is a constant and $f_e = f_e(\mathbf{u})$ does not depend on f. In general, in the nonlinear case one has

$$f = f(t, \mathbf{u}) : \qquad \frac{\partial f}{\partial t} = \nu[f](f - f_e[f]) \,. \tag{1.2.12}$$

What has to be realized is that, sometimes, the complexity of the physical system is so deep that the above modelling is the only way that apparently is left to acquire a complete model.

1.3 Generalized Models and Plan of the Book

Suitable definitions of generalized models, and some preliminary modelling criteria, have been outlined in the preceding section. It was shown that a phenomenological derivation of generalized kinetic equations requires the definition of a state variable and the modelling of possible interactions between the individuals of a large population. In this section the contents of the following chapters will be described. The description will be concise and limited to a few hints concerning the selection of the state variable and the interactions modelling. This will only be introductory to the various class of models that will be discussed in the remaining part of the book, and there described in details.

The contents of Chapter 2 is technical. It consists in a brief presentation of some of the mathematical methods that are used when dealing with generalized models. In particular, some elements of ***probability theory, generalized functions, Lebesgue integration***, and ***weak convergence methods*** are given.

The first class of models is dealt with in Chapter 3, where several models of ***population dynamics with kinetic type interactions*** are presented. They can be regarded as generalizations of the model by Jager and Segel [JAa]. The state variable refers to the social population state of anonymous interacting individuals, and the interactions are assumed to be able to modify only the probability distribution of individuals over their social state. Microscopic interactions are modelled by introducing the two terms we mentioned in the previous section: the ***encounter rate*** and the ***transition probability density***. This class of models can be applied to the analysis of several interesting physical systems, in particular those that refer to social behaviors either within a unique homogeneous population or among several interacting populations. The same model can be used to describe the spread of epidemics over interacting populations.

In Chapter 4 the modelling concerns systems of ***dispersed clusters and droplets***. Here, the interactions can generate either condensation or fragmentation phenomena, and phase transitions may also be taken into account during the evolution, see [PEa]. Despite the great difficulty in dealing with this type of phenomena, it is certainly worthwhile developing research activity towards such a direction due to the great interest for this type of systems in technological applications.

Chapter 5 deals with models on the ***competition between immune system and aggressive hosts***, see [ADa]. Interactions are stated at a cellular level and can either modify the cellular activation state or produce cell proliferation or depletion. In more details, interactions at first modify cellular activation, then cell proliferation and destruction occur as consequences of cell encounters and activation. The model is able to predict the spread of aggressive hosts (say HIV viruses or tumor cells) over the cell environments of vertebrates. We remark that in these models the number of individuals evolve in time. This peculiarity of the evolution generates several interesting mathematical problems including some bifurcation phenomena which will be discussed in the chapter.

The class of models presented in Chapter 6 refers to a well defined physical system, concerning in particular the evolution of ***antigens in the immune system***. Again, a system of population dynamics. The antigens are characterized by the so-called generalized shapes, i.e., geometrical shapes continuously modified by a scalar variable.

Interactions may change the generalized shape; necrosis and inlet from bone marrow rule over death and birth. This class of models, originally proposed by Segel and Perelson [SEb], has the merit of throwing some light on the fascinating world of the immune system. Further developments of this class can hopefully lead to simulate the fundamental behavior of the immune system. The general framework is modelling cellular phenomena in biology [SEa].

Models dealt with in Chapters 3, 4 and 6 are derived by modelling microscopic interactions. On the contrary, the class of models proposed in Chapter 5 consists in an evolution equation that contains phenomenologic terms only indirectly related to microscopic interactions. The differences between these two classes of models are remarked at the end of Section 1.2. Though, a further important analytic difference separates the class of models dealt with in Chapter 3 and those of Chapters 4–6. Indeed, according to Jager and Segel model the number of individuals remains constant with time, whereas in the models of Chapters 4–6 the number of individuals evolves with time. This feature involves, as we shall see, additional difficulties concerning the qualitative analysis of evolution problems.

Chapter 7 is devoted to the ***Boltzmann equation*** and deals with its phenomenological derivation, and to its analytic and computational treatment [BLb], [GLa] and [MAa]. This chapter has to be regarded as a concise guide to the literature on this topic, that is appropriately documented in several specialized books.

Although Boltzmann's is the fundamental model in nonequilibrium statistical mechanics, the various models presented in Chapters 3–6 are not strictly related to it. These models can either be independently derived, or somewhat related to the Jager and Segel model. That is why, although several analogies of some interest can be found to connect them to Boltzmann's, the analysis of this last model has been postponed, also in view of the analytic and computational treatment of the related mathematical problems.

Conversely, the models proposed in Chapters 8 and 9 are more closely related to the Boltzmann equation. This is mostly due to the fact that the state variable $\mathbf{u}$ includes both position and velocity of the elements of the population: in this case physical particles.

The link between models of the first part of the book and models of the second part is not only formal. It has several mathematical and physical implications that are worth to be analyzed in details. In addition, the analysis of those models that are obtained by linking various features of the two different classes above seems to be important. Indeed it may provide perspectives towards new applications.

Chapter 8 deals with ***kinetic models for traffic flows on roads*** or in a network of roads, see [PRa]. The distribution function refers to the number of vehicles contained in the unit tract, while interactions among vehicles may follow conservation equations similar to those of classical mechanics. Vehicles can be modelled by point masses, as in the case of the Boltzmann equation, or by elements of finite dimensions, as in the Enskog equation case. The inlet and outlet of vehicles needs to be taken into account, and this is an additional difficulty which has to be related to the mathematical statement of the problem and to its computational treatment. On the other hand, these problems generally have only a one-dimensional space variable and bounded velocities, which is a simplification with respect to the Boltzmann model.

Chapter 9 still refers to models very close to the Boltzmann equation: they are constituted by ***systems of particles*** assumed as ***point masses*** or ***spheres***. However, the interactions are not modelled in the framework of classical mechanics with conservative interactions. Instead, ***dissipative interactions*** are taken into account. The field of interest of this type of models is that one of applied physics, e.g., semiconductor devices [MKa] and astrophysics [AAa]. In contrast with the Boltzmann equation, each model is based on a different (though more detailed) description of the microscopic interactions. For instance, when the finite dimension of the particles is taken into account together with dissipative collisions, then a dissipative Enskog type model is deduced [ENa], and so on. These models, although derived in the framework of classical mechanics, do not necessarily refer to a gas of particles. For instance, as documented in [AAa], an interesting application is the description of the Saturn rings planets.

Chapter 10 deals with a concise presentation of some details and research perspectives selected among the above topics and mostly related to the personal bias of the authors. Indeed, the reader may develop his own interpretations of the matter and, hence, his personal research line. In details, the chapter deals with the design of discrete models, with the research of a unified general structure for kinetic models, and finally with the analysis of a few research perspectives. By discrete models we mean evolution kinetic equations such that the state variable can attain only a finite number of values. By general structure we mean a structure able to include the largest possible variety of kinetic models.

A collective bibliography of books and review papers is reported at the end of the book with indication of the chapters containing the citations. This will allow the interested reader to link the survey literature to the topics of each chapter.

It is plain that the contents of this book does not cover the whole set of conceivable generalizations that are available, or hidden, in the literature. For instance, a relevant topic such as quantum kinetic theory is not discussed here. Indeed, a conveniently detailed treatment of this topic, starting from the classical paper [MKa], would require much larger size than that one reserved for each chapter in this book. However, the analysis here provided about the various classes of models should be sufficient to indicate a general methodological approach, so that a substantial basis for further developments can be constructed.

1.4 Aim of the Book

After the preceding concise presentation of its contents, it is now possible to go back to the description of the aims of the book. The general idea that inspired it was to develop, in the framework of kinetic models, an analysis similar to that of classical phenomenologic mathematical physics. This idea is introduced in [BLc], where the general framework is given, and explained with the help of a large number of applications.

The traditional approach, developed at the macroscopic scale, provides evolution equations that are in general partial differential equations, and describes physical phenomena in the fields of mechanics, biology, technology etc. However, these models often hide what happens at the molecular scale. Instead, microscopic behaviors are more carefully analyzed by models closely related to the materials structure.

On the other hand, modelling is mostly developed in mechanics. Starting from there, models have been generalized to various other fields of applied sciences to obtain descriptions that may be of use also in different applications. In tracing this procedure, one should consider that no macroscopic approach is possible if the microscopic view is missing in the description of the physical system. This is the gap that the modelling we called ***generalized kinetic modelling*** should fill or, at least, help to reduce.

The interest of this type of modelling in applied sciences is documented in the review paper [BLc] and in the recent collection of surveys [BLd], which reports ten lectures delivered by specialists in the field.

Aim of this book is to provide a general methodological approach to generalized kinetic modelling. As in the case of classical mathematical physics, the starting point can be identified with a few fundamental models. In our case the Jager and Segel model on one hand and the classical Boltzmann equation on the other. From them, several generalizations and developments are possible and will be proposed.

After designing any of these models, the analysis develops along traditional paths: statement of the consequent mathematical problems, qualitative analysis, simulations. For all of them, the mathematical structure consists of a nonlinear integrodifferential equation (rather than a partial differential equation).

Generally a review of the applications is given. This means dealing with simulation problems and hence with the technical solutions of initial and initial-boundary value problems for an equation of the Boltzmann type. This topic, i.e., how to develop methods to compute the solutions, though relevant, is not dealt with in this book, mainly addresses to modelling and analytic treatment of the problems. Solutions to the mathematical problems are mostly obtained by natural developments (often simplifications) of the computational methods available for the Boltzmann equation. They are, in fact, solution schemes nowadays efficiently applied in a large variety of interesting problems in technology, as is reviewed in [NEa].

Examples of the above procedure are proposed, or reviewed, in various chapters of the book. Some of them have to be regarded as only introductory to the various topics, most of which are still waiting for further and deeper analysis.

1.5 References

[AAa] ARAKI S. and TREMAINE S., The dynamics of dense particle disks, *Icarus*, **65** (1986), 83–109.

[ADa] ADAM J.A. and BELLOMO N. Eds., **A Survey of Models on Tumor Immune Systems Dynamics**, Birkhäuser, Boston, (1996).

[BLa] BELLOMO N., LACHOWICZ M., POLEWCZAK J., and TOSCANI G., **Mathematical Topics in Nonlinear Kinetic Theory II: The Enskog Equation**, World Scientific, London, Singapore, (1991).

[BLb] BELLOMO N. Ed., **Lecture Notes on the Mathematical Theory of the Boltzmann Equation**, World Scientific, London, Singapore, (1995).

[BLc] BELLOMO N. and LO SCHIAVO M., From the Boltzmann equation to generalized kinetic models in applied sciences, *Mathl. Comp. Modelling*, **26** (1997), 43–76.

[BLd] BELLOMO N. and PULVIRENTI M. Eds., **Modelling in Applied Sciences: A Kinetic Theory Approach**, Birkhäuser, Boston, (1999).

[CEa] CERCIGNANI C., ILLNER R., and PULVIRENTI M., **Theory and Application of the Boltzmann Equation**, Springer, Berlin, Heidelberg, (1993).

[ENa] ENSKOG D., Kinetiske Theorie, Svenka Akad., **63** (1921). English Translation in **Kinetic Theory**, Brush S. Ed., Vol. 3, Pergamon Press, New York, (1972).

[GLa] GLASSEY R., **The Cauchy Problem in Kinetic Theory**, SIAM Publ., Philadelphia, (1995).

[JAa] JAGER E. and SEGEL L., On the distribution of dominance in a population of interacting anonymous organisms, *SIAM J. Appl. Math.*, **52** (1992), 1442–1468.

[LAa] LACHOWICZ M., Asymptotic analysis of nonlinear kinetic equations: The hydrodynamic limit, in **Lecture Notes on the Mathematical Theory of the Boltzmann Equation** , Bellomo N. Ed., World Scientific, London, Singapore, (1995), 65–148.

[MAa] MASLOVA N., **Nonlinear Evolution Equations**, World Scientific, London, Singapore, (1993).

[MKa] MARKOWICH A., RINGHOFER C., and SCHMEISER C., **Semiconductor Equations**, Springer, Berlin, Heidelberg, (1990).

[NEa] NEUNZERT H. and STRUCKMEIER J., Particle Methods for the Boltzmann equation, in **Acta Numerica 1995**, Cambridge Univ. Press, (1995), 417–458.

[PEa] PENROSE O., LEBOWITZ J., MARRO J., and KALOS M., Growth of clusters in a first-order phase transition, *J. Statist. Phys.*, **19** (1978), 243–267.

[PRa] PRIGOGINE I. and HERMAN R., **Kinetic Theory of Vehicular Traffic**, Elsevier, New York, (1971).

[SEa] SEGEL L., **Modelling Dynamic Phenomena in Molecular and Cellular Biology**, Cambridge Univ. Press, (1984).

[SEb] SEGEL L. A. and PERELSON A. S., Computations in shape space: A new approach to immune network theory, in **Theoretical Immunology II, SFI Studies in the Science of Complexity**, Perelson A. S. Ed., Addison-Wesley, Reading, (1988), 321–342.

[TRa] TRUESDELL C. and MUNCASTER R., **Fundamentals of Maxwell Kinetic Theory of a Simple Monoatomic Gas**, Academic Press, New York, (1980).

Chapter 2

Mathematical Background: Measure, Integration, Topology

2.1 Introduction

This chapter briefly reports about some of the mathematical tools that constitute the background of the contents developed in the chapters that follow. This will be done not only for future reference, but also to collect a short and not-applied summary of those formal aspects and theoretical methods that are connected with the Boltzmann equation, and that may be extended to other kinetic equations of the Boltzmann type developed in more general frameworks.

Indeed, most of the mathematical methods used in kinetic theory are not restricted to the world of physical particles, that represents the starting point of Boltzmann equation, but on the contrary, they can be successfully used in many other fields of physical and social sciences. The mathematics being the same, it will be kept here as general as possible, and only from time to time some connections will be made with the original strictly kinetic world just for the matter of examples.

In this chapter we do not develop any new theory. On the contrary, we only summarize the main theoretical results that are going to be used in the following part of this book. Therefore, these arguments are extracted from classical books, which we refer to for additional details, without further discussions. This will also force us to recall, with no comment, well known properties and definitions.

The chapter is divided into four sections:

Section 2.1 is this introduction.

Section 2.2 deals with the basic notions necessary to introduce measures and density functions. A particular emphasis is on the real spaces, which are the ones that are used in the following, and hence to the definition of Stieltjes measures in $n \geq 1$ dimensional real spaces.

Section 2.3 sketches the fundamentals and properties of Lebesgue integrals. Few theorems of integration theory are also mentioned. Then, some of the connections with distribution theory are enlightened.

Section 2.4 is an introduction to the use of weak convergence methods and compactness methods such as Ascoli-Arzelà theorem.

When necessary, each section is further divided to better distinguish the various arguments which, in such a fast review, are unavoidably tightly condensed.

Actually, some of the topics reported in this chapter are not of immediate use for the analysis of generalized kinetic models. However, the whole contents aims to constitute the necessary background for any deeper insight into the related mathematical analysis with special attention to qualitative theory. The reader who already possesses a sufficient knowledge on the above topics may simply skip this chapter.

2.2 Tools from Measure Theory

The mathematical models developed in the chapters that follow consist of evolution equations wherein the dependent variable is a probability density or a number density function. Probability theory is the basic background for the above models. This section deals with probability measures and random variables.

2.2.1 Physical connections

A physical (microscopical) ***system*** of particles is characterized by its ***states***, each modelled by a point x in an n-dimensional real vector space $\mathbf{R}^n$ equipped by some convenient norm. This is the base ***phase space***. On the phase space it may be defined, or at least assumed to exist, a family of ***evolution operators*** T indexed by a real variable t that takes values in a connected subset I of $\mathbb{R}$:

$$T_{t_0,t_1} : \mathbb{R}^n \to \mathbb{R}^n, \qquad t_0, t_1 \in I\,.$$

The evolution operators are linked with the solutions of, and possibly are determined by, a set of evolution equations, which in general are intrinsically deterministic and time-reversal invariant. Consequently the operators are such that

$$T_{t_0,t_2} = T_{t_0,t_1} \circ T_{t_1,t_2}\,,$$

when both the latter exist, and

$$T_{t_0,t_1} \circ T_{t_1,t_0} = \mathbb{1}\,,$$

when any of the two exist. Furthermore, each of the operators T_{t_0,t_1} is differentiable, invertible, and with a differentiable inverse on $\mathbb{R}^n$.

For example, T_{t_0,t_1} may be the solution of an initial value problem of an ordinary differential equation such as

$$\dot{x} = v(t,x), \qquad x(t_0) = \tilde{x}_0\,, \tag{2.2.1}$$

where $v : \mathbf{R} \times \mathbb{R}^n \to \mathbb{R}^n$ is sufficiently smooth to guarantee well-posedness of the problem. Possibly, t is allowed to take only discrete values; then T_{t_0,t_1} satisfies a discrete equation such as

$$x_{j+1} = g(x_j), \qquad x_0 = \tilde{x}_0\,, \tag{2.2.2}$$

where $g : \mathbb{R}^n \times \mathbb{R}^n$ is a map of $\mathbb{R}^n$ in itself.

In the case of a classical conservative system of N physical particles, sometimes called ***microcanonical ensemble***, the phase space is $\mathbb{R}^{6N}$ and the defining equations are the Hamilton's autonomous divergenceless equations of motion. In this case (and for sufficiently regular fields) the family of evolution operators is a one-parameter local group of diffeomorphisms from $\mathbb{R}^{6N}$ to $\mathbb{R}^{6N}$ (or, better, from the manifold $\mathcal{H}^{-1}(\mathcal{E}) \subset \mathbb{R}^{6N}$ to itself).

When the number n is large, or when physical considerations do not allow the complete knowledge of the point x in the phase space, or when the mathematics involved is by far too difficult to be treated, the exact procedure may be accomplished only in principle, and other methods are needed to obtain results of interest. On the other hand, in these cases the results that are looked for are not necessarily as detailed as the complete knowledge of the evolution

$$x(t, x_0) := T_{t_0,t}\, x_0 \,.$$

On the contrary, they may be limited to assign values to the so-called ***macroscopic*** or ***observable*** variables, which are only loosely connected with each single state. In fact these values do depend on the underlying evolution T, however the dependence is averaged over certain ***sets of states*** that the particles, or the system, may presumably assume at the given instant. In this case, what is called a (statistical) ***state*** of the system is the function itself that assigns a common weight to a whole set of (microscopical) states.

A probabilistic structure is thus added to the phase space, and even if the underlying dynamical system remains strictly deterministic, the observable quantities assume the meaning of random variables, and their evolutions that of stochastic processes. Here we recall some of the main aspects of this procedure, which may be found, among others, in [ASa], [HAa], [KOa], [LOa], and [ROa].

2.2.2 Basics on measure theory

A set A of points in the space $\mathbf{X}$ is called an ***event*** when it is suitable to represent a set of microscopical states, any one of which may be assumed by the given system at a given instant of time. Hence, events are subsets of the whole space, which is certainty.

On the other hand, some assumptions are necessary in order that the model be realistic and may represent the results of actual physical observations. This yields to require a structure, on the family of events, which may correspond to that one of the experimental measurements on the system.

A convenient requirement for a family of subsets of $2^{\mathbf{X}}$ to be a family of events $\mathcal{S} \subset 2^{\mathbf{X}}$ is that it satisfies the following:

Definition 2.2.1. ***σ-algebra*** *is a family $\mathcal{S} \subset 2^{\mathbf{X}}$ of subsets of $\mathbf{X}$ such that:*

a) $\emptyset \in \mathcal{S}$ *and* $\mathbf{X} \in \mathcal{S}$;
b) if $S \in \mathcal{S}$ *then* $\mathbf{X} \setminus S \in \mathcal{S}$;
c) if $S_1, S_2, \ldots,$ *belong to* $\mathcal{S}$ *then* $\bigcup_{i=1}^{\infty} S_i$ *belongs to* $\mathcal{S}$.

Any subset $\mathcal{F} \subset 2^{\mathbf{X}}$ can be used as a starting family to ***generate*** a (minimal) σ-algebra $\mathcal{S}_{\mathcal{F}}$ containing $\mathcal{F}$. This may be done, for instance, by collecting all the intersections and countable unions of sets starting from those in $\mathcal{F}$. A notable example is the ***Borel*** σ-algebra $\mathcal{B}(\mathbf{X})$ which is the smallest σ-algebra containing the open sets of $\mathbf{X}$.

Sometimes even simpler families are introduced.

Definition 2.2.2. ***Ring*** *of sets is a family $\mathcal{R} \subset 2^{\mathbf{X}}$ which contains the empty set and is closed under finite intersections and symmetric differences.*

Definition 2.2.3. ***Algebra*** *of sets is a family $\mathcal{A} \subset 2^{\mathbf{X}}$ which is a ring and contains the space $\mathbf{X}$.*

Well known relations of set theory imply then that rings and algebras are closed under finite unions and differences of sets. Hence, an algebra still has properties a) and b) above, whereas c) holds true only for finite families $\{S_i \in \mathcal{A}\}_{i=1,\dots,j}$, $j \in \mathbb{N}$. The following are also of use:

Definition 2.2.4. ***Semi-ring*** *on* $\mathbf{X}$ *is a family* $\mathcal{F} \subset 2^{\mathbf{X}}$ *which contains the empty set, it is closed under finite intersections, and:*

c) if F *belongs to* $\mathcal{F}$, *and* $F_1 \subset F$ *belongs to* $\mathcal{F}\setminus\{\emptyset\}$, *then a finite family* $\{F_i\}_{i=2,\dots,j}$ *of pairwise disjoint sets in* $\mathcal{F}$ *may be found such that* $F = \bigcup_{i=1}^{j} F_i$.

Definition 2.2.5. ***Semi-algebra*** *is a semi-ring* $\mathcal{S}_0$ *containing the whole space* $\mathbf{X}$.

For instance the family of semiclosed intervals $(a,b]$ (see later) is a semi-ring in $\mathbb{R}^n$ and a semi-algebra in $\overline{\mathbb{R}}^n$. Here and in the following we shall use the notation $\overline{\mathbb{R}} := \mathbb{R} \cup \{-\infty\} \cup \{+\infty\}$, $\mathbb{R}_+ := \{x \in \mathbb{R} \,|\, x \geq 0\}$, and $\overline{\mathbb{R}}_+ := \mathbb{R}_+ \cup \{+\infty\}$.

Any element S of a σ-algebra $\mathcal{S}$ is by definition a ***physically plausible*** event. The reason is that, now, a (non-negative) number may be assigned in connection with, or even to represent, the set altogether. In particular, a number may be assigned to the actual occurrence of any of the states which belong to the set, or about the expected value of a state function.

Definition 2.2.6. ***Measure*** *on* $(\mathbf{X}, \mathcal{S})$ *is a set function* $\mu : \mathcal{S} \to \overline{\mathbb{R}}_+$ *such that*

a) μ *is* ***non-negative***, *and* $\mu(\emptyset) = 0$;

b) μ *is* ***subadditive***, *i.e., if* S *and* S' *belong to* $\mathcal{S}$ *then* $\mu(S \cup S') \leq \mu(S) + \mu(S')$;

c) μ *is* σ***-additive***, *i.e., if* $S_i \in \mathcal{S}$ *for* $i = 1, 2, \dots$, *and if* $S_i \cap S_j = \emptyset$ *for* $i \neq j$ *then*

$$\mu\left(\bigcup_{i=1}^{\infty} S_i\right) = \sum_{i=1}^{\infty} \mu(S_i). \tag{2.2.3}$$

The measure is called ***finitely additive*** if the last property is granted only for finite collections of sets in $\mathcal{S}$.

Definition 2.2.7. ***Measure space*** *is a triple* $(\mathbf{X}, \mathcal{S}, \mu)$.

Definition 2.2.8. *A measure* μ*, or in general a* σ*-additive set function* $m : \mathcal{F} \subseteq 2^{\mathbf{X}} \to \overline{\mathbb{R}}_+$*, is said to be* ***$\sigma$-finite on X*** *when* $\mathbf{X}$ *may be covered by a countable collection of sets* $F_i \in \mathcal{F}$, $i = 1, 2, \ldots$*, all having* $m(F_i) < \infty$.

Usually, measures μ are constructed by extending some given σ-additive set function $m : \mathcal{F} \to \overline{\mathbb{R}}_+$ defined on a semi-ring (or a semi-algebra) $\mathcal{F}$, to the σ-algebra $\mathcal{S}_{\mathcal{F}}$ generated by $\mathcal{F}$. The construction may be done by means of the following procedure if further assumption is made that the function m is σ-finite.

a) A subadditive set function μ^* is defined on $2^{\mathbf{X}}$ such that for each set $A \in 2^{\mathbf{X}}$ one has

$$\mu^*(A) := \inf \left\{ \sum_{j=1}^{\infty} m(F_j) \right\}; \quad F_j \in \mathcal{F}, \quad A \subseteq \bigcup_{j=1}^{\infty} F_j . \qquad (2.2.4)$$

b) A set $E \in 2^{\mathbf{X}}$ is defined: ***measurable with respect to*** μ^* when for each $A \in 2^{\mathbf{X}}$ one has

$$\mu^*(A) = \mu^*(A \backslash E) + \mu^*(A \cap E) . \qquad (2.2.5)$$

In general, any set S of a σ-algebra $\mathcal{S}$ is called ***measurable.***

c) The collection $\mathcal{M}$ of measurable set is a σ-algebra (direct sum of σ-algebras) containing $\mathcal{F}$ and hence containing $\mathcal{S}_{\mathcal{F}}$.

d) The function μ^* is a measure on $\mathcal{M}$, hence so is the restriction of μ^* to $\mathcal{S}_{\mathcal{F}}$; i.e., one defines $\bar{\mu}(S) := \mu^*(S)$ for each $S \in \mathcal{S}_{\mathcal{F}}$.

e) If necessary, $\bar{\mu}$ may be ***completed*** in μ by defining $\mu(S \cup N) := \bar{\mu}(S)$ for $S \in \mathcal{S}$, $N \subseteq \widetilde{N}$ and $\widetilde{N} \in \mathcal{S}$ such that $\bar{\mu}(\widetilde{N}) = 0$.

f) The complete measure μ proves to be unique, σ-additive and σ-finite. (With a different procedure this last assumption may be released.)

σ-additivity is generally accepted as a fundamental property of measures on account of the fact that: σ-additive set functions $m : \mathcal{F} \to \overline{\mathbb{R}}$ are continuous. Indeed:

a) if $F_1, F_2, \ldots \in \mathcal{F}$ are such that $F_j \uparrow F$ then

$$\lim_{j\to\infty} m(F_j) = m(F);$$

b) if $F_1, F_2, \ldots \in \mathcal{F}$ are such that $F_j \downarrow F$, and $m(F_1) < \infty$, then

$$\lim_{j\to\infty} m(F_j) = m(F).$$

The converse is true if additivity and either one of the two properties are assumed.

Among the σ-algebras that may be introduced on a real space (which is what we are interested in) the most used and well known is the following.

Definition 2.2.9. ***Borel σ-algebra*** *$\mathcal{B}(\overline{\mathbb{R}}^n)$ is the smallest σ-algebra of subsets of $\overline{\mathbb{R}}^n$ containing the open sets in the natural topology generated by the spheres induced by the Euclidean metric.*

Hence, $\mathcal{B}(\overline{\mathbb{R}}^n)$ is the σ-algebra generated by the semialgebra of the semiclosed intervals $(a, b]$.

Definition 2.2.10. *$\mathcal{S}$**-measurable** (or: **measurable**) is said of a real function $f : (\mathbf{X}, \mathcal{S}, \mu) \to \overline{\mathbb{R}}$ such that the set $f^{-1}(B)$ belongs to $\mathcal{S}$ whenever B belongs to $\mathcal{B}(\overline{\mathbb{R}})$.*

Definition 2.2.11. ***Random variable*** *on $(\mathbf{X}, \mathcal{S}, \mu)$ is an $\mathcal{S}$-measurable real function in probability theory (i.e., when $\mu(\mathbf{X}) < \infty$).*

In this way a number may be assigned to the occurrence of the event: ***the observed value $r(x)$ of the random variable r belongs to a certain neighborhood of a certain value $r^* \in \overline{\mathbb{R}}$.***

It is notable that measurability is closed under sums, products, functional composition (when $\mathcal{S} \equiv \mathcal{B}(\overline{\mathbf{R}})$) and pointwise convergence. Moreover, measurability is extended from a function to another when the two are almost equal, where:

Definition 2.2.12. ***Almost true*** *is said of a property on a measure space $(\mathbf{X}, \mathcal{S}, \mu)$ when the property is true on $\mathbf{X}$ except, at most, a set of points of $\mathbf{X}$ which (is measurable and) has measure zero.*

It may be shown that if $f, g : \mathbf{X} \to \overline{\mathbf{R}}$ are defined on a set $S \in \mathcal{S}$ and $f = g$ a.e. (that is: f equals g almost everywhere) on S then f is measurable whenever g is.

The concept of ***almost everywhere*** may be used to introduce special kind of convergence.

Definition 2.2.13. *A sequence $\{f_j\}_{j\in\mathbf{N}}$ of measurable functions $f_j : (\mathbf{X}, \mathcal{S}, \mu) \to \mathbf{R}$ is said to be* ***almost everywhere convergent*** *to a function f when the set of points where the convergence fails has measure zero.*

Definition 2.2.14. *A sequence $\{f_j\}_{j\in\mathbf{N}}$ of measurable functions $f_j : (\mathbf{X}, \mathcal{S}, \mu) \to \mathbf{R}$ is said to be* ***almost uniformly convergent*** *to a function f when for any $\delta > 0$ there exists $S_\delta \in \mathcal{S}$ such that $\mu(\mathbf{X} \setminus S_\delta) < \delta$ and $\lim_{j\to\infty} f_j = f$ uniformly on S_δ.*

Definition 2.2.15. *A sequence $\{f_j\}_{j\in\mathbf{N}}$ of measurable functions $f_j : (\mathbf{X}, \mathcal{S}, \mu) \to \mathbf{R}$ is said to be* ***convergent in measure*** *to a (measurable) function f when, for each $\epsilon > 0$, if the following is satisfied:*

$$\lim_{j\to\infty} \mu\left\{x \,\middle|\, |f_j(x) - f(x)| \geq \epsilon\right\} = 0\,. \tag{2.2.6}$$

An a.e. convergent sequence of measurable functions f_j has a measurable limit. If a sequence of functions $\{f_j\}_{j\in\mathbf{N}}$ is a.e. convergent on a set S that has $\mu(S) < \infty$ then the convergence is almost uniform. Convergence in measure only implies existence of an a.e. convergent subsequence $\{f_{j_k}\}_{k\in\mathbf{N}} \subset \{f_j\}_{j\in\mathbf{N}}$ to the same function f.

On finite measure spaces convergence in measure is implied by a.e. convergence since uniform convergence implies convergence a.e. and in measure.

2.2.3 Measures on the real line

A measure which is by far the most used on real spaces is the following

Definition 2.2.16. ***Lebesgue measure*** λ *is the (completion of the) measure induced on* $\mathcal{B}(\overline{\mathbf{R}}^n)$ *by the function which assigns to each interval* $(a,b] \subset \mathbf{R}$ *its length:*

$$\lambda((a,b]) = b - a\,. \tag{2.2.7}$$

Hence to the semiclosed intervals $(a^1, b^1] \times (a^2, b^2] \times \cdots \times (a^n, b^n]$ the product is given by $\prod_{i=1}^{n}(b^i - a^i)$. Moreover: $\lambda(a,b] \equiv \lambda[a,b] \equiv \lambda(a,b) := \lambda(a,b]$. If $a, b = \pm\infty$ then $\lambda = +\infty$. In addition, λ is σ-additive and σ-finite.

Clearly, λ is not the only useful measure on $\mathcal{B}(\overline{\mathbf{R}}^n)$, and several convenient generalizations are possible. Let us begin with $n = 1$.

Definition 2.2.17. ***Lebesgue-Stieltjes measure****, or* $\mathcal{LS}$*- measure, is any measure on* $\mathcal{B}(\overline{\mathbf{R}})$ *which is finite on the bounded intervals.*

Definition 2.2.18. ***Generating distribution function*** *on* $\mathbf{R}$ *is a function* $s : \mathbf{R} \to \mathbf{R}$ *such that*

a) *s is **non-decreasing**; i.e., $a, b \in \mathbf{R}$, and $a < b$ imply $s(a) \leq s(b)$;*
b) *s is **right-continuous**; i.e., $a_j > a$, for $j = 1, 2, \ldots$, and $a_j \downarrow a$ imply $\lim_{j\to\infty} s(a_j) = s(a)$.*

Any generating distribution function $s : \mathbf{R} \to \mathbf{R}$ induces a unique $\mathcal{LS}$-measure μ_s on $\mathcal{B}(\overline{\mathbf{R}})$. Indeed, by defining

$$\mu_s(a,b] := s(b) - s(a)\,, \qquad a, b \in \mathbf{R}, \quad a < b\,, \tag{2.2.8}$$

one obtains an additive function on the semi-ring of finite disjoint unions of right-semiclosed intervals, and since this function is seen to be σ-additive on this field (and possibly σ-finite) then the procedure above ensures that it is possible to define a unique complete measure μ_s on $\mathcal{B}(\overline{\mathbb{R}})$ (σ-finite on $\mathcal{B}(\mathbb{R})$) starting from a generating distribution function s.

The following are useful properties of μ_s.

a) s may be extended (owing to its monotonicity) to become a function $s : \overline{\mathbb{R}} \to \overline{\mathbb{R}}$ by setting:

$$s(-\infty) := \lim_{x \to -\infty} s(x) \quad \text{and} \quad s(+\infty) := \lim_{x \to +\infty} s(x) ; \qquad (2.2.9)$$

hence the numbers a, b in Eq. (2.2.8) are accepted to be $\pm\infty$, and formula (2.2.8) defines a σ-additive measure on $\mathcal{B}(\overline{\mathbb{R}})$;

b) if $a_j \downarrow a$ then $\mu_s(a, a_j] \to 0$ and $\mu_s(a_j, b] \to \mu_s(a, b]$;

c) using the identity $(a, b) = \bigcup_{j=1}^{\infty}(a, b_j]$ if $b_j \uparrow b$, the following are easily seen:

$$\mu_s(a, b) = \lim_{j \to \infty} \mu_s(a, b_j] = s(b^-) - s(a) ;$$

$$\mu_s[a, b] = s(b) - s(a^-) = \mu_s\{a\} + \mu_s(a, b] ;$$

$$\mu_s[a, b) = s(b^-) - s(a^-) = \mu_s\{a\} + \mu_s(a, b] ;$$

$$\mu_s[-\infty, b] = \mu_s(-\infty, b] = s(b) - s(-\infty) ;$$

d) $\mu_s\{a\} = 0$ iff s is continuous at a; $\mu_s\{a\} = s(a) - s(a^-)$, otherwise;

e) Lebesgue measure λ is the induced measure μ_s when $s(x) = x$.

Conversely, any $\mathcal{LS}$-measure μ on $\mathcal{B}(\overline{\mathbb{R}})$ defines, up to an arbitrary constant (say $s(a)$), a unique generating distribution function s, for instance:

$$s(x) := s(a) + \mu(a, x] \qquad \text{if} \quad x > a,$$

$$s(x) := s(a) - \mu(x, a] \qquad \text{if} \quad x < 0.$$

If, say, $\mu(-\infty, a] < \infty$, then it may be set $s(x) := s(-\infty) + \mu(-\infty, x]$.

Easy and particular examples of $\mathcal{LS}$-measures are obtained when the generating distribution function $s \equiv s_c$ is defined by use of a density function, where:

Definition 2.2.19. ***Density function*** *$f : \mathbb{R} \to \mathbb{R}_+$ is a non-negative locally (Riemann or Lebesgue) integrable function, see Definition 2.3.4.*

Example 2.2.1. A continuous generating distribution function $s_c : \mathbb{R} \to \mathbb{R}$ is found if one starts from an integrable function $f : \mathbb{R} \to \mathbb{R}$ and sets

$$s_c(x) := s_c(a) + \int_a^x f(\xi)d\xi \,. \tag{2.2.10}$$

(In fact, s_c is absolutely continuous and hence a.e. differentiable.) For instance, Lebesgue measure λ has $f(\xi) = 1$.

□

Example 2.2.2. A discontinuous generating distribution function $s_d : \mathbb{R} \to \overline{\mathbb{R}}$ is, for instance, the following stepwise constant function:

$$s_d(x) = s_d(x_i)\,, \quad \text{for} \quad x_i \le x < x_{i+1} \quad \text{and} \quad x_1, x_2, \ldots \in \mathbb{R}\,,$$

where the set $\mathcal{D}_s := \{x_1, x_2, \ldots\}$ is finite or countable and, for example, it has been assumed the existence of a ***first*** point x_1 such that $s_d(-\infty) = s_d(x_1^-) =: h_0 \ge 0$. With the notation

$$h_i := s_d(x_i) - s_d(x_i^-) = s_d(x_{i+1}^-) - s_d(x_i^-) \ge 0\,, \qquad i = 1, 2, \ldots\,,$$

one has

$$s_d(x_i) = \sum_{k=0}^{k=i} h_k\,, \qquad \text{for} \quad i = 1, 2, \ldots\,,$$

and the generated measure is discrete and satisfies

$$\mu_{s_d}\{x\} = 0 \quad \text{for} \quad x \neq x_i, \qquad \mu_{s_d}\{x_i\} = h_i, \quad i = 1, 2, \dots .$$

It follows that:

$$\mu_{s_d}(S) = \sum_{x_i \in S} h_i ,$$

where the sum is done over all the x_i's in D_s that belong to S. The set D_s may even contain convergent subsequences $\{x_{i1}, x_{i2}, \dots\}$ such that

$$\lim_{k\to\infty} x_{ik} = x_{i\infty} \in D_s$$

provided that

$$\sum_{k=1}^{\infty} h_{ik} < \infty .$$

□

If one introduces the

Definition 2.2.20. ***Heaviside function*** $H_a : \mathbb{R} \to \mathbb{R}$, *is the function*

$$H_a(x) = 0 \qquad \text{if} \quad x < a, \qquad \text{and} \qquad H_a(x) = 1 \qquad \text{if} \quad x \geq a ,$$

then the above generating distribution function s_d may be written as

$$s_d(x) = h_0 + \sum_{i=1}^{\infty} h_i H_{x_i}(x) . \tag{2.2.11}$$

Example 2.2.3. A singular generating distribution function s_s is a continuous function, monotone, hence of bounded variation, not absolutely continuous, a.e. differentiable, although with $s_s' = 0$ where defined. A singular generating distribution function verifies the formula

$$s_s(x) - s_s(a) = \mu_{s_s}(a, x]$$

where μ_{s_s} is an $\mathcal{LS}$-measure such that $\mu_{s_s}\{x\} = 0$ for each $x \in \mathbb{R}$, yet there exists a set N of Lebesgue measure zero such that $\mu_{s_s}(\mathbb{R} \setminus N) = 0$. The devil's staircase, or Cantor function on $[0,1]$, is a singular generating distribution function.

□

Since any monotone function $s : \mathbb{R} \to \mathbb{R}$ may be uniquely separated into the sum of its three components: absolutely continuous, discrete and singular, it follows that any generating distribution function on $\mathbb{R}$ induces an $\mathcal{LS}$-measure that may be correspondingly decomposed into its three components μ_{s_c}, μ_{s_d}, μ_{s_s} and, conversely, any $\mathcal{LS}$-measure μ_s on $\mathbb{R}$ may be uniquely decomposed into the sum $\mu_s = \mu_{s_c} + \mu_{s_d} + \mu_{s_s}$. It may also be seen that any measure on $\mathbb{R}$ is an $\mathcal{LS}$-measure.

If a measure μ is already defined on $\mathbf{X}$, and $r : \mathbf{X} \to \overline{\mathbb{R}}$ is a real valued measurable function, i.e., a random variable, then the following may be defined:

Definition 2.2.21. ***Marginal measure***, *relative to* r, *or:* ***measure induced by*** r, *is the measure* μ_r *on* $(\overline{\mathbb{R}}, \mathcal{B}(\overline{\mathbb{R}}))$ *defined by*

$$\mu_r(B) := \mu\left(r^{-1}(B)\right), \qquad B \in \mathcal{B}(\overline{\mathbb{R}}). \tag{2.2.12}$$

Provided that $\mu\left(r^{-1}((-\infty, a])\right) < \infty$ for at least one $a \in \mathbb{R}$, one may also introduce

Definition 2.2.22. ***Generating distribution function relative to*** r *is the function* $s_r : \mathbb{R} \to \mathbb{R}_+$ *defined by*

$$s_r(x) := \mu_r(-\infty, x] := \mu\left(r^{-1}((-\infty, x])\right), \quad x \in \mathbb{R}. \tag{2.2.13}$$

Indeed, the function s_r is non-decreasing right-continuous, and it has been set $s_r(-\infty) = 0$. One may remark that

$$s_r(b) - s_r(a) =: \mu_{s_r}(a, b] = \mu_r(a, b] \,.$$

It may also be remarked that μ_r is defined on the image of r whereas μ_s is defined on the domain of s. For instance, if the function

$$r = \sum_{i=1}^{\infty} a_i \chi_{S_i}$$

is elementary (see later) then its correspondent distribution function s_r is discontinuous (see Example 2.2.2):

$$s_r(x) = \sum_{i=1}^{\infty} h_i H_{x_i}(x) \,, \quad \text{and} \quad \mu_{s_r}(S) = \sum_{x_i \in S} \mu(S_i) \,,$$

where $x_i \equiv a_i$ and $h_i := \mu_{s_r}\{x_i\} \equiv \mu(S_i) \,, \quad i = 1, 2, \ldots.$

2.2.4 Measures on $\mathbf{R}^n$

When the dimension of the space $\mathbf{R}^n$ is greater than one, firstly the concepts of interval and of $\mathcal{LS}$-measure are to be extended.

Definition 2.2.23. ***Intervals*** $(a, b]$ *on* $\mathbf{R}^n$ *are defined by*

$$(a, b] := \left\{ x \in \mathbf{R}^n \,\middle|\, x^i \in (a^i, b^i] \,, \quad i = 1, \ldots, n \,, \quad a \le b \right\} \tag{2.2.14}$$

where the superscripts are the indices of the components, and where $a \le b$ *means* $a^i \le b^i$ *for all* $i = 1, \ldots, n$.

In particular one has

$$(a, +\infty) := \left\{ x \in \mathbf{R}^n \,\middle|\, x^i > a^i \,, \quad i = 1, \ldots, n \right\} \,, \tag{2.2.15}$$

and

$$(-\infty, b] := \left\{x \in \mathbb{R}^n \,\middle|\, x^i \leq b^i, \quad i = 1, \ldots, n\right\}. \tag{2.2.16}$$

Definition 2.2.24. *$\mathcal{LS}$-**measure** μ on $\mathbb{R}^n$ is a measure defined on $\mathcal{B}(\overline{\mathbb{R}}^n)$ which is finite on any bounded interval of $(\mathbb{R}^n)$.*

Starting from a $\mathcal{LS}$-measure μ one may define a non-negative function $s : \mathbb{R}^n \to \overline{\mathbb{R}}_+$, (assume that $\mu(-\infty, x] < \infty$ for at least an $x \in \mathbb{R}^n$), by means of

$$s(x) := \mu_s(-\infty, x], \qquad x \in \mathbb{R}^n. \tag{2.2.17}$$

However, some care has now to be taken since the difference $s(b) - s(a)$ may not give, even if $a \leq b$, the correct value for the measure $\mu(a, b]$; indeed, it is not generally true that

$$(-\infty, b] = (-\infty, a] \cup (a, b], \qquad a \leq b, \qquad a, b \in \mathbb{R}^n. \tag{2.2.18}$$

To relate the interval $(a, b]$ to the semi-infinite intervals $(-\infty, x]$, the so-called ***difference operators*** may be used, defined by (provided $a \leq b$)

$$(a, b] \equiv (a^1, \ldots, a^n; b^1, \ldots, b^n] =: \Delta_{a^1,b^1}\Delta_{a^2,b^2} \ldots \Delta_{a^n,b^n}(-\infty,\ x]$$
$$= \Delta_{a^1,b^1}\left[\Delta_{a^2,b^2}\left[\cdots\left[\Delta_{a^{n-1},b^{n-1}}\left[\Delta_{a^n,b^n}(-\infty,\ x]\right]\right]\cdots\right]\right],$$

where

$$\Delta_{a^i,b^i}[S(x)] := S(x^1, \ldots, x^{i-1}, b^i, x^{i+1}, \ldots, x^n)$$
$$\setminus S(x^1, \ldots, x^{i-1}, a^i, x^{i+1}, \ldots, x^n).$$

For instance, one has

$$\begin{aligned}&\Delta_{a^{n-1},b^{n-1}}\left[\Delta_{a^n,b^n}(-\infty\,;x]\right]\\&=\left[(-\infty\,;x^1,\dots,b^{n-1},b^n]\setminus(-\infty\,;x^1,\dots,b^{n-1},a^n]\right]\\&\setminus\left[(-\infty;\,x^1,\dots,a^{n-1},b^n]\setminus(-\infty;\,x^1,\dots,a^{n-1},a^n]\right].\end{aligned}$$

Correspondingly, the following are introduced.

Let $s : \mathbb{R}^n \to \overline{\mathbb{R}}$ be a function, and $a, b, x \in \mathbb{R}^n$. For $j = n-1, \dots, 1$, let the functions $s^{(j)}_{a;b} : \mathbb{R}^j \to \overline{\mathbb{R}}$ be recursively given by

$$\begin{aligned}s^{(n-1)}_{a;b} &:= \Delta_{a^n,b^n}\, s(x) = s(x^1,\dots,x^{n-1},b^n) - s(x^1,\dots,x^{n-1},a^n);\\ \dots\quad &\dots\quad \dots\\ s^{(j)}_{a;b} &:= \Delta_{a^{j+1},b^{j+1}} s^{(j+1)}_{a;b}(x^1,\dots,x^{j+1})\\ &= s^{(j+1)}_{a;b}(x^1,\dots,x^j,b^{j+1}) - s^{(j+1)}_{a;b}(x^1,\dots,x^j,a^{j+1});\\ \dots\quad &\dots\quad \dots\\ s^{(1)}_{a;b} &:= \Delta_{a^2,b^2} s^{(2)}_{a;b}(x^1,x^2) = s^{(2)}_{a;b}(x^1,b^2) - s^{(2)}_{a;b}(x^1,a^2).\end{aligned}$$

Finally, let $s^{(0)}_{a;b} \in \mathbb{R}$ be given by

$$\begin{aligned}s^{(0)}_{a;b} &:= \Delta_{a^1,b^1}\Delta_{a^2,b^2}\cdots\Delta_{a^n,b^n}\, s(x)\\ &= \Delta_{a^1,b^1} s^{(1)}_{a;b}(x^1) = s^{(1)}_{a;b}(b^1) - s^{(1)}_{a;b}(a^1).\end{aligned}\tag{2.2.19}$$

Definition 2.2.25. *A function $s : \mathbb{R}^n \to \overline{\mathbb{R}}$ is said to be **non-decreasing on** $\mathbb{R}^n$ if for $a, b \in \overline{\mathbb{R}}^n$, $a \le b$, it happens that $s^{(0)}_{a;b}$ (as well as all the functions $s^{(j)}_{a;b}$ for $j = 1, \dots, n-1$) defined above are non-negative.*

Definition 2.2.26. *A function $s : \mathbb{R}^n \to \overline{\mathbb{R}}$ is said to be **right-continuous on** $\mathbb{R}^n$ if $x_k \in \mathbb{R}^n$, $x_k \downarrow x$, $k = 1, 2, \dots,$ imply that $\lim_{k\to\infty} s(x_k) = s(x)$.*

Definition 2.2.27. ***Generating distribution function on*** $\mathbb{R}^n$ *is a non-decreasing right continuous extended function* $s : \mathbb{R}^n \to \overline{\mathbb{R}}$.

Definition 2.2.28. ***k-marginal generating distribution functions*** $s^{(k)} : \mathbb{R}^k \to \mathbb{R}$ *of a generating distribution function* $s : \mathbb{R}^n \to \overline{\mathbb{R}}$, *are the functions defined for* $1 \le k \le n-1$, *by*

$$s^{(k)} := s^{(k)}_{-\infty;+\infty}(x^{i_1}, \dots, x^{i_k}) \equiv \lim_{\xi \to \infty} s^{(k+1)}(x^{i_1}, \dots, x^{i_k}, \xi),$$

where $(i_1, \dots, i_k)$ *is a subset of* $(1, \dots, n)$, *and* $\xi \in \mathbb{R}^{n-k}$ *is the vector that refers to the remaining* $(n-k)$ *coordinates.*

(Clearly, if the domain of the generating distribution function is a compact interval $[\alpha, \beta] \subset \overline{\mathbb{R}}^n$, then the above limits are to be computed for ξ approaching the correspondent subvector of β.

As well as in $\mathbb{R}$, an arbitrary generating distribution function $s : \mathbb{R}^n \to \overline{\mathbb{R}}$ uniquely defines a $\mathcal{LS}$-measure μ_s on $\mathcal{B}(\overline{\mathbb{R}}^n)$ by means of (2.2.19) and

$$\mu_s(a, b] := s^{(0)}_{a;b} \equiv \Delta_{a^1,b^1} \Delta_{a^2,b^2} \cdots \Delta_{a^n,b^n} \, s(x). \tag{2.2.20}$$

Conversely one may see that a finite $\mathcal{LS}$-measure μ correctly defines, via formula (2.2.17), a generating distribution function $s : \mathbb{R}^n \to \overline{\mathbb{R}}$ such that $\mu(a, b] = s^{(0)}_{a;b}$. If $(-\infty, b]$ is not finite, then $\mu(a, b]$ may be used only to define $s^{(0)}_{a;b}$ and formula (2.2.19) has to be traced backwards.

Notable examples of generating distribution functions on $\mathbb{R}^n$, and consequently of $\mathcal{LS}$-measures on $\mathcal{B}(\overline{\mathbb{R}}^n)$ are obtained as follows.

Example 2.2.4. If $s^1, \dots, s^n$ are generating distribution functions in $\mathbb{R}$, then the function $s(x) := \prod_{i=1}^n s^i(x^i)$ is a generating distribution function in $\mathbb{R}^n$ and

$$\mu_s(a, b] = \prod_{i=1}^n \left(s^i(b^i) - s^i(a^i) \right). \tag{2.2.21}$$

In particular, the Lebesgue measure on $\mathcal{B}(\overline{\mathbb{R}}^n)$ is given by

$$\lambda(a, b] = \prod_{i=1}^{n} (b^i - a^i) . \tag{2.2.22}$$

□

Example 2.2.5. If $f : \mathbb{R}^n \to \mathbb{R}_+$ is a non-negative Riemann integrable function, i.e., there exists finite Riemann-integral:

$$\int_{-\infty}^{+\infty} \cdots \int_{-\infty}^{+\infty} f(x^1, \ldots, x^n) dx^1 \cdots dx^n , \tag{2.2.23}$$

then for all $a, b \in \overline{\mathbb{R}}^n$ such that $a \leq b$, it is possible to set

$$\begin{aligned} s(x) &:= \int_{-\infty}^{x^n} \cdots \int_{-\infty}^{x^1} f(\xi^1, \ldots, \xi^n) d\xi^1 \cdots d\xi^n \\ &=: \mathcal{R} \int_{(-\infty, x]} f(\xi) d\xi , \end{aligned} \tag{2.2.24}$$

and

$$\begin{aligned} \mu_s(a, b] &:= \int_{a^n}^{b^n} \cdots \int_{a^1}^{b^1} f(x^1, \ldots, x^n) dx^1 \cdots dx^n \\ &=: \mathcal{R} \int_{(a,b]} f(x) dx . \end{aligned} \tag{2.2.25}$$

It is notable that in this way an absolutely continuous function $s : \overline{\mathbb{R}}^n \to \mathbb{R}_+$ is obtained. This procedure is going to be extended in the following paragraph.

2.3 Tools from Integration Theory

In nearly all models that are presented in this book, the macroscopic observable quantities are obtained as suitable averages of microscopic quantities by means of convenient probability density functions. This implies that one has to carefully handle the subjects of probabilities and of integration.

The aim of the present section is to fill the gap between the concept of (abstract) probability density function and that of macroscopically measurable physical observable.

2.3.1 Lebesgue integrals

The last example seen in the previous section may be considerably generalized using the following fundamental tools.

Definition 2.3.1. ***Indicator function*** *or* ***characteristic function*** *of a set B in $2^{\mathbf{X}}$ is a function $\chi_B : \mathbf{X} \to \mathbf{R}$ such that $\chi_B(x) = 1$ when $x \in B$, and zero otherwise.*

Definition 2.3.2. ***Simple function*** *is a function $\sigma : (\mathbf{X}, \mathcal{S}) \to \mathbb{R}$ defined by*

$$\sigma(x) = \sum_{i=1}^{i=m_\sigma} a_i \chi_{S_i}(x) , \qquad a_j \in \mathbf{R} , \qquad j = 1, 2, \ldots, m_\sigma \in \mathbf{N} , \tag{2.3.1}$$

where $S_1, S_2, \ldots, S_{m_\sigma} \in \mathcal{S}$ is a finite family of disjoint sets in $\mathcal{S}$.

The function σ is called ***elementary*** if the family $S_1, S_2, \ldots$ is allowed to be countable.

Let $(\mathbf{X}, \mathcal{S}, \mu)$ be a measure space, and $\mathbf{M}(\mathcal{S}, \mathcal{B}(\mathbb{R}))$ be the set of all $\mathcal{S}$-measurable functions $f : \mathbf{X} \to \mathbb{R}$. The family $\mathbf{M}$ certainly contains both the continuous functions $f : \mathbf{X} \to \mathbf{R}$, and the indicators χ_S of exactly those sets S which belong to $\mathcal{S}$.

Let $\mathcal{F}_s$ be the family of all simple functions

$$\sigma = \sum_{i=1}^{i=m_\sigma} a_i \chi_{S_i}$$

from $(\mathbf{X}, \mathcal{S})$ to $\mathbb{R}$. We recall ([HAa] §21) that: every measurable function $f : \mathbf{X} \to \overline{\mathbb{R}}_+$ is the (pointwise) limit of a sequence of simple functions σ_n. In addition, if f is non-negative then the simple functions may be assumed to be non-negative and the sequence increasing. The limit is uniform if the functions σ_n are elementary.

Definition 2.3.3. ***Lebesgue integral*** *of a non-negative measurable function* $f : \mathbf{X} \to \overline{\mathbb{R}}_+$ *is defined by*

$$\int_{\mathbf{X}} f d\mu := \sup \left\{ \sum_{j=1}^{m_\sigma} a_j \mu(S_j) \right\} \leq \infty , \qquad (2.3.2)$$

where the supremum is done over all $\sigma \in \mathcal{F}_s$ *such that* $\sigma(x) \leq f(x)$.

Definition 2.3.4. ***Lebesgue integrable function*** $f : \mathbf{X} \to \overline{\mathbb{R}}$ *is a measurable function* f *such that its positive and negative parts:* f_+ *and* f_- *(where* $f_\pm(x) := \max(\pm f(x), 0)$*) both have finite Lebesgue integrals. Equivalently: a measurable* $f : \mathbf{X} \to \overline{\mathbb{R}}$ *is integrable when both* f_+ *and* f_- *admit a sequence of elementary functions, whose integrals are finite, that is uniformly convergent to* f_+ *and* f_- *respectively.*

A measurable function f is integrable if and only if $|f|$ is integrable. When $(\mathbf{X}, \mathcal{S}, \mu)$ is σ-finite, the integrability condition is: that all the integrals

$$\int_{E_n} |f| d\mu$$

be finite for each μ-finite covering $\{E_i\}_{i\in\mathbb{N}}$ of $\mathbf{X} = \bigcup_{i=1}^{\infty} E_i$, and that

$$\lim_{j\to\infty} \int_{\bigcup_{i=1}^{j} E_i} f d\mu$$

be finite and independent of the covering.

Definition 2.3.5. ***Lebesgue integral on*** *$S \in \mathcal{S}$ of a measurable function $f : \mathbf{X} \to \overline{\mathbb{R}}$ is defined by*

$$\int_S f d\mu := \int_{\mathbf{X}} f_+ \, \chi_S \, d\mu - \int_{\mathbf{X}} f_- \, \chi_S \, d\mu , \tag{2.3.3}$$

unless both the positive and negative parts of f have infinite integrals.

Lebesgue integrals are linear monotone functionals on the set $\mathbf{M}(\mathcal{S},\mathcal{B}(\overline{\mathbb{R}}))$ of the measurable functions $f : (\mathbf{X},\mathcal{S}) \to \overline{\mathbb{R}}$; moreover, they are σ-additive, and continuous with respect to the measure μ. Finally, a non-negative function f has zero Lebesgue integral iff $f = 0$ μ-a.e.

2.3.2 Main theorems of integration theory

The last property has a self-standing importance in that it allows a further, and most used, kind of convergence. The following are well known results.

In the space $\mathbf{M}\left(\mathcal{S},\mathcal{B}(\overline{\mathbb{R}})\right)$ of measurable functions $f : (\mathbf{X},\mathcal{S}) \to \overline{\mathbb{R}}$, let $\sim$ be the equivalence relation defined by $f_1 \sim f_2$ μ-a.e., and call $\widetilde{\mathbf{M}} := \mathbf{M}/\sim$ the space of equivalence classes in $\mathbf{M}$.

Definition 2.3.6. *$L_p(\mathbf{X},\mathcal{S},\mu)$, or briefly L_p, for $p \in [1,\infty)$, is the space of the classes $\widetilde{f}$ in $\widetilde{\mathbf{M}}$ that satisfy*

$$\int_{\mathbf{X}} |\widetilde{f}|^p \, d\mu < \infty .$$

In the following, as usual, the superscript will be ignored and instead of ***class*** the word ***function*** will be used.

$$\|\cdot\|_p := \left(\int_{\mathbf{X}} |f|^p \, d\mu\right)^{1/p} \quad \text{is a norm on} \quad L_p .$$

Definition 2.3.7. *$L_\infty(\mathbf{X}, \mathcal{S}, \mu)$, or briefly L_∞, is the space of μ-a.e. bounded functions $f : \mathbf{X} \to \overline{\mathbb{R}}$, equipped with the norm*

$$\|f\|_\infty := \operatorname{ess\,sup} |f(x)| := \inf_{f \in \tilde{f}} \sup_{t \in \mathbb{R}} f(t)$$

$$\equiv \inf \left\{ \alpha \in \overline{\mathbb{R}} \,\middle|\, \lambda \{x \mid f(x) > \alpha\} = 0 \right\} .$$

Definition 2.3.8. ***Convergent in the mean****, or simply* ***convergent in*** *L_p, is a sequence $\{f_j\}_{j \in \mathbb{N}}$ of (equivalence classes of) functions $f_j \in L_p$, with $p \in [1, \infty)$, such that*

$$\lim_{j \to \infty} \|f - f_j\|_p = 0 .$$

Convergence in $(L_\infty, \|\cdot\|_\infty)$ is the uniform a.e. convergence.

THEOREM 2.3.1. ***(Riesz-Fisher)*** *The spaces $\left(L_p, \|\cdot\|_p\right)$, for $p \in [1, \infty]$, are complete with respect to their norms (or* ***Banach spaces****) meaning that any Cauchy sequence in L_p converges in the p-mean to a unique point of L_p.*

THEOREM 2.3.2. ***(Hölder inequality)*** *Let $p, q \in [1, \infty]$ be such that $p^{-1} + q^{-1} = 1$. Then*

$$\int_{\mathbf{X}} |fg| \, d\mu \;\le\; \|f\|_p \cdot \|g\|_q , \qquad f \in L_p , \quad g \in L_q . \tag{2.3.4}$$

The following results (see, for instance, [HAa]) allow one to obtain the integral of a measurable function f as the limit of the integrals of a sequence of functions $\{f_j\}_{j\in\mathbb{N}}$ that converges to f. Recall at first the notation:

$$\underline{\lim}\,\xi_j \equiv \liminf \xi_j := \sup_{j\in\mathbb{N}} \inf_{k\geq j} \xi_k\,, \tag{2.3.5}$$

where $\{\xi_j\}_{j\in\mathbb{N}}$ is a sequence of points in a metric space Ξ.

THEOREM 2.3.3. ***(Fatou's Lemma)*** *Let $\{f_j\}_{j\in\mathbb{N}}$ be a sequence of non-negative integrable functions $f_j : (\mathbf{X}, \mathcal{S}, \mu) \to \overline{\mathbb{R}}_+$ such that*

$$\underline{\lim} \int_{\mathbf{X}} f_j d\mu < \infty\,,$$

then the function $f : (\mathbf{X}, \mathcal{S}, \mu) \to \overline{\mathbb{R}}_+$, pointwise defined by $f(x) := \underline{\lim}\, f_j(x)$, is integrable and

$$\int_{\mathbf{X}} f d\mu \leq \underline{\lim} \int_{\mathbf{X}} f_j d\mu\,.$$

THEOREM 2.3.4. ***(Monotone Convergence)*** *Let $\{f_j\}_{j\in\mathbb{N}}$ be a non-decreasing sequence of extended non-negative measurable functions $f_j : (\mathbf{X}, \mathcal{S}, \mu) \to \overline{\mathbb{R}}_+$ which converges a.e. on $S \in \mathcal{S}$ to a function f. Then*

$$\lim_{j\to\infty} \int_S f_j d\mu = \int_S f d\mu\,. \tag{2.3.6}$$

THEOREM 2.3.5. ***(Dominated Convergence)*** *Let $\{f_j\}_{j\in\mathbb{N}}$ be a sequence of measurable functions $f_j : (\mathbf{X}, \mathcal{S}, \mu) \to \overline{\mathbb{R}}$ which converges a.e. (or in measure) on $S \in \mathcal{S}$ to a function f. Let $g : \mathbf{X} \to \overline{\mathbb{R}}_+$ be an integrable function over $S \in \mathcal{S}$ such that $|f_j(x)| \leq g(x)$ for all $j \in \mathbb{N}$ and a.e. on S. Then $f_1, f_2 \ldots$ and*

f are integrable on S and $\{f_j\}_{j\in\mathbb{N}}$ converges in the mean (hence in measure) to f. Consequently

$$\lim_{j\to\infty}\int_S f_j d\mu = \int_S f d\mu .$$

THEOREM 2.3.6. ***(Monotone Convergence in L_1, B. Levi)*** *Let $\{f_j\}_{j\in\mathbb{N}}$ be a non-decreasing sequence of integrable functions $f_j : (\mathbf{X}, \mathcal{S}, \mu) \to \overline{\mathbb{R}}_+$ with uniformly bounded integrals. Then the sequence converges a.e. to a (finite) integrable function f and*

$$\lim_{j\to\infty}\int_S f_j d\mu = \int_S f d\mu .$$

Corollary: Let $\{f_j\}_{j\in\mathbb{N}}$ be a sequence of measurable functions $f_j : (\mathbf{X}, \mathcal{S}, \mu) \to \overline{\mathbb{R}}$ such that

$$\sum_{j=1}^{\infty}\int_{\mathbf{X}} |f_j| d\mu < \infty .$$

Then the sequence

$$\left\{\sum_{j=1}^{j=k} f_j\right\}_{k\in\mathbb{N}}$$

converges a.e. to a finite valued function $\sum_{j=1}^{\infty} f_j$ such that

$$\sum_{j=1}^{\infty}\int_{\mathbf{X}} f_j d\mu = \int_{\mathbf{X}}\left(\sum_{j=1}^{\infty} f_j\right) d\mu . \tag{2.3.7}$$

In the case $(\mathbf{X}, \mathcal{S}, \mu) \equiv (\mathbb{R}^n, \mathcal{B}(\mathbb{R}^n), \lambda)$ the above theorems give the following well known results ([ASa] Section 1.6.).

Let $a, b, c, d \in \overline{\mathbb{R}}$ and $f : (a,b) \times (c,d) \to \mathbb{R}$. Assume that for each $x \in (a,b)$ the function $f(x,\cdot) : (c,d) \to \mathbb{R}$ is measurable. Then:

a) If for (almost) all $x \in (a,b)$ the function $|f(x,\cdot)|$ is bounded by an integrable function $g : (c,d) \to \mathbb{R}$ and if $x_0 \in (a,b)$ is such that $\lim_{x \to x_0} f(x,y)$ exists for (almost) each $y \in (c,d)$ then

$$\lim_{x \to x_0} \int_a^b f(x,y)dy = \int_a^b \left[\lim_{x \to x_0} f(x,y) \right] dy \,. \tag{2.3.8}$$

b) Assume that $f(x,\cdot)$ is integrable over (c,d) and that $\frac{\partial}{\partial x} f$ exists on all $(a,b) \times (c,d)$. If for all $x \in (a,b)$ the function $\left|\frac{\partial}{\partial x} f(x,\cdot)\right|$ is bounded by an integrable function $g : (c,d) \to \mathbb{R}$ then for each $x \in (a,b)$ the following exist and are equal

$$\frac{d}{dx} \int_a^b f(x,y)dy = \int_a^b \frac{\partial}{\partial x} f(x,y)dy \,. \tag{2.3.9}$$

Again in the case $(\mathbf{X}, \mathcal{S}, \mu) \equiv (\mathbb{R}^n, \mathcal{B}(\mathbb{R}^n), \lambda)$, it may be useful to briefly compare Riemann integration and Lebesgue integration. One has

$$\int_{[a,b]} f d\lambda = \mathcal{R} \int_{[a,b]} f(x)dx \,, \qquad a, b, x \in \mathbb{R}^n \,, \tag{2.3.10}$$

whenever the ***proper Riemann integral*** on the right-hand side exists (finite), i.e., there exists, finite and independent of the partition of $[a,b]$, a unique limit value for the upper and lower sums of the function f when the partition is arbitrarily refined. The Riemann integral exists ([ASa] Section 1.7) if and only if f is bounded and λ-a.e. continuous on $[a,b]$.

On the other hand, sometimes an a.e. continuous function $f : B \to \mathbb{R}$, where B denotes an interval $[a,b] \subseteq \overline{\mathbb{R}}^n$, may still be considered to be Riemann-integrable even if it does not have (a finite) limit when approaching (a subset of) ∂B, or if B is unbounded.

Definition 2.3.9. ***Improper Riemann integral*** *is defined as*

$$\mathcal{R}\int_B f(x)dx := \lim_{j\to\infty} \mathcal{R}\int_{B_j} f(x)dx\,, \tag{2.3.11}$$

whenever for all sequences $\{B_j\}_{j\in\mathbf{N}}$ *of intervals* $B_j := [a_j, b_j] \subset B$, *such that* $B_j \uparrow B$ *and* $\lambda(B_j) < \infty$, *it happens that the limit* $\lim_{j\to\infty} \mathcal{R}\int_{B_j} f(x)dx$ *exists and is finite and independent of the choice of the sequence* $\{B_j\}_{j\in\mathbf{N}}$.

In these cases, the corresponding Lebesgue integral may not exist. It does exist, and it is equal to the Riemann integral if, in addition, the absolute value $|f|$ has (possibly improper) Riemann integral ([KOa] Section 30.3).

Furthermore, when $(\mathbf{X}, \mathcal{S}, \mu) \equiv (\mathbb{R}^n, \mathcal{B}(\mathbb{R}^n), \mu_s)$ and μ_s is an $\mathcal{LS}$-measure corresponding to a generating distribution function $s : \mathbb{R}^n \to \mathbb{R}$, the following may be introduced.

Definition 2.3.10. ***Lebesgue-Stieltjes integral***

$$\int_B f\,d\mu_s \equiv \int_B f(x)\,ds(x)$$

is the Lebesgue integral defined by use of the Borel sets and of an $\mathcal{LS}$*-measure* μ_s.

Definition 2.3.11. ***Riemann-Stieltjes integral*** $\mathcal{R}\int f(x)ds(x)$ *is defined by use of the common limit value, if it exists and is independent of the partition of* $[a, b]$ *e.g in intervals* $(x_{i-1}, x_i]$, *of the partial sums obtained by use of the products* $M_j \cdot \mu_s(x_{j-1}, x_j]$ *and* $m_j \cdot \mu_s(x_{j-1}, x_j]$, *where* M_j *and* m_j *are respectively the superior and inferior values of* f *on the interval* $(x_{j-1}, x_j] \subset \mathbb{R}^n$, *and* $\bigcup(x_{j-1}, x_j] = [a, b]$ *(with the obvious closures on the boundary).*

It may be shown ([KOa] Section 36.4) that for $f \in C^0([a, b])$ the two integrals exist and are equal.

Example 2.3.1. If s_c is a generating distribution function on $\mathbb{R}$ such that

$$s_c = \int_{-\infty}^{x} s'_c(\xi)d\xi\,,$$

then

$$\int_{\mathbb{R}} f d\mu_{s_c} = \int_{-\infty}^{+\infty} f(\xi)ds_c(\xi) = \int_{-\infty}^{+\infty} f(\xi)s'_c(\xi)d\xi\,.$$

□

Example 2.3.2. A discrete Stieltjes measure generalizes what happens when a real valued random vector may assume only a (countable) set of values $x_1, x_2, \ldots$ with probabilities $p_1, p_2, \ldots$.

A discrete Stieltjes measure is a measure μ_{s_d} concentrated on a countable set of points $\mathcal{D}_s := \{x_i\}_{i\in\mathbf{Z}} \subset \mathbb{R}$ (see Example 2.2.2):

$$\mu_{s_d}(S) = \sum_{x_i \in S} \mu_{s_d}\{x_i\}\,, \quad \text{and} \quad \mu_{s_d}\{x_i\} = h_i \geq 0\,, \quad i \in \mathbf{Z}\,.$$

Under the assumption, for instance, that there exists a first point x_1 such that $\mu_{s_d}\{x\} = 0$ for all $x < x_i$, the measure μ_{s_d} may correspond to a stepwise constant generating distribution function such as

$$s_d(x) = h_0 + \sum_{i=1}^{\infty} h_i H_{x_i}(x)\,.$$

The $\mathcal{LS}$-integral of a function $f : \mathbb{R}^n \to \mathbb{R}$ with respect to this measure reduces to

$$\int_{\mathbb{R}} f d\mu_{s_d} = \sum_{i=1}^{\infty} f(x_i)h_i\,.$$

□

By recalling that any function $\Phi : \mathbf{R} \to \mathbf{R}$ of bounded variation may be represented by the difference $\Phi = s_1 - s_2$ of two monotone functions, one may extend the concept of $\mathcal{LS}$ integral of a function $f : \mathbf{R} \to \mathbf{R}$ and define

$$\int_B f(x)\, d\Phi(x) := \int_B f(x)\, ds_1(x) - \int_B f(x)\, ds_2(x)\,.$$

In this sense, a particular version ([KOa] Section 36.6) of Riesz theorem 2.4.3 states that any continuous linear functional on $C^0([a,b])$ may be represented by a Stieltjes integral.

In the following chapters, as well as in the theories of probability and statistical mechanics, Lebesgue and Lebesgue-Stieltjes integrals are used unless explicitly stated.

2.3.3 Distribution functionals

The examples seen above have many useful consequences if the family is conveniently restricted of functions f on which the $\mathcal{LS}$-integral is considered. Here we recall some of the main definitions on the subject.

Definition 2.3.12. ***Space* T *of test functions*** *is the linear space of the functions* $\varphi : \mathbf{R}^n \to \mathbf{R}$ *defined on compact support* $K_\varphi \subset \mathbf{R}^n$ *and such that on* K_φ *there exist continuous partial derivatives of all orders* $k = 0, 1, \ldots$

$$\varphi^{(\mathbf{k})} := \frac{\partial^k \varphi}{\partial^{\nu_1} x^1 \cdots \partial^{\nu_n} x^n}\,,$$

where

$$\mathbf{k} := (\nu_1, \ldots, \nu_n)\,, \qquad k := |\mathbf{k}| := \sum_{i=1}^{n} \nu_i\,, \qquad \nu_1, \ldots, \nu_n \in \mathbf{N}\,.$$

Definition 2.3.13. *A sequence* $\{\varphi_j\}_{j\in\mathbf{N}}$ ***converges*** *to* $\varphi \in \mathbf{T}$ *when: there exists a compact* $K_\varphi \subset \mathbf{R}^n$ *such that* $K_\varphi \supseteq K_{\varphi_j}$ *for all*

$j \in \mathbf{N}$ and that the sequence $\left\{\varphi_j^{(\mathbf{k})}\right\}_{j\in\mathbf{N}}$ converges to $\varphi^{(\mathbf{k})}$ uniformly on K_φ and for all $k = 0, 1, \ldots,$, and all $\mathbf{k}$ such that $|\mathbf{k}| = k$.

$\mathbf{T}$ may be seen as the union $\mathbf{T} = \bigcup_{m=1}^{\infty} \mathbf{T}_m$ of test function spaces such that the functions in $\mathbf{T}_m$ have supports contained in the closed spheres: $\left\{x \in \mathbb{R}^n \,\middle|\, |x| \le m\right\}$, and ([KOa] Section 21) the topologies on $\mathbf{T}_m$ are generated by the subbase of the neighborhoods of the origin

$$B_\epsilon^{m,j}(0) = \left\{\varphi \in \mathbf{T}_m \,\middle|\, \|\varphi\|_j < \epsilon\right\} \quad \subset \; B_\epsilon^{m,j-1}(0)\,,$$

where $j = 1, 2, \ldots,\quad \epsilon > 0$, and

$$\|\varphi\|_j = \sup_{|x|\le m,\ |\mathbf{k}|\le j} \left|\varphi^{(\mathbf{k})}(x)\right|, \quad j = 1, 2, \ldots .$$

Definition 2.3.14. ***Space*** $\mathbf{D}(\mathbf{T})$ ***of the distributions on*** $\mathbf{T}$ *is the linear space of the continuous linear functionals $\widehat{g} : \mathbf{T} \to \mathbb{R}$; meaning that*

a) $\widehat{g}(c_1\varphi_1 + c_2\varphi_2) = c_1\widehat{g}(\varphi_1) + c_2\widehat{g}(\varphi_2)\,;\quad c_1, c_2 \in \mathbb{R};\quad \varphi_1, \varphi_2 \in \mathbf{T}$,

b) $\widehat{g}(\lim_{j\to\infty}\varphi_j) = \lim_{j\to\infty}\widehat{g}(\varphi_j)\,,$ *if* $\lim_{j\to\infty}\varphi_j$ *belongs to* $\mathbf{T}$.

Example 2.3.3. If $f : \mathbb{R}^n \to \mathbb{R}$ is integrable on compact sets and $\varphi \in \mathbf{T}$ then the following integral exists and belongs to $\mathbf{D}(\mathbf{T})$

$$\widehat{f}(\varphi) := \int_{\mathbb{R}^n} \varphi \; d\mu_{s_c} \equiv \int_{\mathbb{R}^n} f\,\varphi\; d\lambda\,;$$

here the $\mathcal{LS}$-measure is generated by a distribution $s_c : \mathbb{R}^n \to \mathbb{R}$ such that

$$s_c(x) = \int_{(-\infty,x]} f \; d\lambda\,.$$

□

Example 2.3.4. Consider a point-discrete $\mathcal{LS}$-measure:

$$\mu_{s_d}(S) = \sum_{x_i \in S} \mu_{s_d}\{x_i\} ,$$

with stepwise constant generating distribution function

$$s_d(x) = \sum_{i=1}^{\infty} h_i H_{x_i}(x) .$$

Then the $\mathcal{LS}$-integral

$$\widehat{h}(\varphi) := \int_{\mathbf{R}^n} \varphi d\mu_{s_d} = \sum_{i=1}^{\infty} \varphi(x_i) h_i$$

defines a continuous linear functional on $\mathbf{T}$, and hence defines a point in $\mathbf{D}(\mathbf{T})$.

□

It is customary to denote this functional with the help of the so-called ***delta functions*** which directly express the property of μ_{s_d} to be concentrated on the points $x_1, x_2, \ldots$; that is, to set

$$\sum_{i=1}^{\infty} h_i \varphi(x_i) = \int_{\mathbf{R}^n} \varphi \, d\mu_{s_d} =: \sum_{i=1}^{\infty} h_i \int_{\mathbf{R}^n} \varphi(x)\delta(x - x_i)\, dx$$

$$=: \sum_{i=1}^{\infty} h_i \widehat{\delta}_{x_i}(\varphi) . \qquad (2.3.12)$$

It is also possible to extend this notation outside of $\mathbf{T}$. For instance in the particular, μ-finite, case

$$h_0 =: \mu_d\{x_0\} = \mu_d(\mathbf{R}^n) = 1 , \qquad x_0 \in \mathbf{R}^n ,$$

one may extend it to $\varphi(x) = 1$ for all $x \in \mathbb{R}^n$ and write, for the measure μ_d,

$$\int_{\mathbb{R}^n} d\mu_d \;=:\; \int_{\mathbb{R}^n} \delta(x - x_0)dx \;=:\; \widehat{\delta}_{x_0}(1) \;=\; 1\,. \tag{2.3.13}$$

More in general, for μ_d defined as above, and $\psi : \mathbb{R}^n \to \mathbb{R}$ continuous at x_0, one may state the following

Definition 2.3.15. *The **(shifted)** δ **function***

$$\delta_{x_0}(x) := \delta(x - x_0), \qquad x_0 \in \mathbb{R}^n\,,$$

is defined by

$$h_0\psi(x_0) \;\equiv\; \int_{\mathbb{R}^n} \psi\, d\mu_d \;=:\; h_0 \int_{\mathbb{R}^n} \psi(x)\, \delta(x - x_0)dx \;=:\; h_0\widehat{\delta}_{x_0}(\psi)\,. \tag{2.3.14}$$

Another possible way to construct the space $\mathbf{D}(\mathbf{T})$ may be recognized in the following steps.

a) ***δ-sequences*** $\{\delta_{,k}\}_{k\in\mathbb{N}}$ are sequences in $\mathbf{T}$ such that
   a1) $\delta_{,k}(x) \geq 0$ for all $x \in \mathbb{R}^n$ and $k \in \mathbb{N}$;
   a2) $\delta_{,k}(x) = 0$ for all x such that $|x| > \frac{1}{k}$;
   a3) $\widehat{\delta}_{,k}(1) := \int_{\mathbb{R}^n} \delta_{,k}(x)dx = 1$ for all $k \in \mathbb{N}$.

b) The ***mean value theorem*** may ([KOa] Section 20.3) be used to see that if $\{\delta_{,k}\}_{k\in\mathbb{N}}$ is a δ-sequence and φ is in $\mathbf{T}$, (or even if φ is continuous with compact support), then

$$\lim_{k\to\infty} \int_{\mathbb{R}^n} \delta_{,k}\, \varphi\, dx \;=\; \varphi(0) \;=:\; \widehat{\delta}(\varphi)\,, \tag{2.3.15}$$

i.e., the following may be written, for all $\varphi \in \mathbf{T}$

$$\lim_{k\to\infty} \int_{\mathbb{R}^n} \delta_{,k}\, \varphi\, dx =: \int_{\mathbb{R}^n} \delta\, \varphi\, dx = \varphi(0) =: \widehat{\delta}(\varphi)\,. \tag{2.3.16}$$

c) convergence ***in the sense of distributions*** is to write the above relation in the following way ([KOa] Section 21.3):

$$\lim_{k\to\infty} \widehat{\delta}_{,k}(\varphi) \quad =: \quad \widehat{\delta}(\varphi)\,. \tag{2.3.17}$$

Definition 2.3.15. *The* ***(shifted)*** δ ***function*** *is*

$$\delta_{x_0}(x) := \delta(x - x_0), \qquad x_0 \in \mathbf{R}^n\,,$$

where δ *is defined by* (2.3.15).

Similarly: a distribution $\widehat{f} \in \mathbf{D}(\mathbf{T}) : \mathbf{T} \to \mathbb{R}$ is defined not only by means of

$$\widehat{f}(\varphi) := \int_{\mathbf{R}^n} f(x)\varphi(x)\, dx\,, \qquad \varphi \in \mathbf{T},$$

when a function $f : \mathbb{R}^n \to \mathbb{R}$ exists (in $L_1(\mathbf{R}^n, \mathcal{B}(\mathbb{R}^n), \lambda)$) such that the $\mathcal{LS}$-integral on the right-hand side is meaningful, but also, as in the δ-case for which such a function is not available, by defining for each $\varphi \in \mathbf{T}$ the functional $\widehat{f}(\varphi)$ as the limit of the sequence $\{\widehat{f}_k(\varphi)\}_{k\in\mathbf{N}} \subset \mathbf{T}$, provided that this limit is linear and continuous in the sense seen above.

Example 2.3.5. ([KOa] Section 21.5) Let $\varphi \in \mathbf{T} : \mathbb{R} \to \mathbb{R}$ be such that $\varphi(x) = 0$ for $x \notin [-m, +m]$ and some $m \in \mathbb{R}$. Then continuity of φ ensures that the following distribution exists and is finite (provided that the integrals are understood in the sense of the Cauchy's principal value)

$$\begin{aligned}\widehat{x^{-1}}(\varphi) &:= \int_{-\infty}^{+\infty} \frac{\varphi(x)}{x}\, dx = \int_{-m}^{m} \frac{(\varphi(x) - \varphi(0))}{x}\, dx \\ &+ \int_{-m}^{m} \frac{\varphi(0)}{x}\, dx = \int_{-m}^{m} \frac{(\varphi(x) - \varphi(0))}{x}\, dx,\end{aligned} \tag{2.3.18}$$

since

$$\int_{-m}^{+m} \frac{\varphi(0)}{x}\, dx = \lim_{k\to\infty} \left(\int_{-m}^{-1/k} \frac{\varphi(0)}{x}\, dx + \int_{1/k}^{+m} \frac{\varphi(0)}{x}\, dx \right) = 0.$$

□

It may also be useful to recall that the space $\mathbf{T}$ separates the points of $C(\mathbb{R}^n, \mathbb{R})$, and, more in general, of $L_1(\mathbb{R}^n, \mathcal{B}(\mathbb{R}^n), \lambda)$.

Example 2.3.6. Let $\psi \in \mathbf{T} : \mathbb{R}^n \to \mathbb{R}$ be such that $\psi(x) > 0$ for $|x| < 1$ and $\psi(x) = 0$ for $|x| \geq 1$. Let $M := \int_{\mathbb{R}^n} \psi dx$ and define

$$\delta_{,1} := \frac{\psi}{M}, \qquad \text{and} \quad \delta_{,k}(x) := k^n\, \psi(kx), \qquad \text{for} \quad k \in \mathbb{N}.$$

The sequence $\{\delta_{,k}\}_{k\in\mathbb{N}}$ is a δ-sequence in $\mathbf{T}$.

□

Computations may be introduced on $\mathbf{D}(\mathbf{T})$.

Let $f : \mathbb{R}^n \to \mathbb{R}$ be integrable on the compact subsets of an open set $V \subseteq \mathbb{R}^n$ and let φ be in $\mathbf{T}$ (or at least φ be continuous and of compact support: $[a_\varphi, b_\varphi]$).

Definition 2.3.16. ***Convolution product*** *of f with φ, is the function $f * \varphi : V_\varphi \subset \mathbb{R}^n \to \mathbb{R}$ defined by*

$$\begin{aligned}(f * \varphi)(x) :&= \int_{[a_\varphi, b_\varphi]} f(x - \xi)\, \varphi(\xi)\, d\xi, \\ &= \int_{[x - b_\varphi, x - a_\varphi]} f(t)\, \varphi(x - t)\, dt,\end{aligned}$$

where

$$x \in V_\varphi; \qquad V_\varphi := \left\{ x \in \mathbb{R}^n \,\middle|\, [x - b_\varphi, x - a_\varphi] \subset V \right\}.$$

Both $-a_\varphi$ and b_φ may be set equal to ∞ with the position that the product is zero as soon as one of the factors is zero.

Example 2.3.7. If $\{\delta_{,k}\}_{k\in\mathbf{N}}$ is a δ-sequence and $f : \mathbb{R}^n \to \mathbb{R}$ is continuous in $V \subset \mathbf{R}^n$ then ([ANa] Section 3.1.1) the sequence $\{f * \delta_{,k}\}_{k\in\mathbf{N}}$ of the C^∞-functions $f * \delta_{,k}$ converges to f uniformly on each compact subset K of V.

□

If φ, f and $\varphi * f$ are as above then differentiation is meaningful, and one has: $(f * \varphi)^{(\mathbf{k})} = (f * \varphi^{(\mathbf{k})})$ on V_φ, for every $\mathbf{k} \in \mathbb{N}^n$.

Example 2.3.8. A distribution may be differentiated as if integration by parts were allowed on the ***function*** f:

$$\frac{\partial^k}{\partial^{\nu_1} x^1 \cdots \partial^{\nu_n} x^n}\widehat{f}(\varphi) = (-1)^k \widehat{f}\left(\frac{\partial^k}{\partial^{\nu_1} x^1 \cdots \partial^{\nu_n} x^n}\varphi\right).$$

It implies that if $\widehat{f}_j \rightharpoonup \widehat{f}$ in the sense of (2.3.17) then $\widehat{f}_j^{(\mathbf{k})} \rightharpoonup \widehat{f}^{(\mathbf{k})}$ for every $|\mathbf{k}| \in \mathbf{N}$.

□

Example 2.3.9. The Heaviside function H_a (e.g., in $\mathbf{R}$) is such that

$$\widehat{H}_a(\varphi) = \int_{-\infty}^{\infty} H_a(x)\varphi(x)dx = \int_a^{\infty} \varphi(x)dx\,,$$

hence it follows that

$$\widehat{H}'_a(\varphi) = -\widehat{H}_a(\varphi') = -\int_a^{\infty} \varphi'(x)dx = \varphi(a)$$

$$= \widehat{\delta}_a(\varphi) = \int_{-\infty}^{\infty} \delta(x-a)\varphi(x)dx\,.$$

For instance:

$$s_d(x) := \sum_{i=1}^{\infty} h_i H_{x_i}(x)$$

yields

$$\widehat{s}'_d(x) = \sum_{i=1}^{\infty} h_i \widehat{\delta}(x - x_i)\,.$$

□

Example 2.3.10. Still in the dim=1 case, for each $\widehat{f} \in \mathbf{D}$ the ordinary differential equation $y' = f(x)$ has a solution (in the sense of distributions) given by

$$\widehat{y}(\varphi) = \widehat{f}\left(-\int_{-\infty}^{x} \varphi(\xi) d\xi\right)\,.$$

□

2.3.4 Some notable generalizations

It is now possible to generalize formula (2.2.25). If $h : (\mathbf{X}, \mathcal{S}, \mu) \to \overline{\mathbb{R}}_+$ is measurable and if, for each $S \in \mathcal{S}$, one defines

$$\nu(S) := \int_S h \; d\mu\,, \tag{2.3.19}$$

then the function $\nu : \mathcal{S} \to \overline{\mathbb{R}}_+$ is a measure on $\mathcal{S}$, possibly completed with all the subsets of μ-null measure; and recall that $0 \cdot \infty = 0$. In the case when $(\mathbf{X}, \mathcal{S}, \mu)$ is σ-finite then ν is σ-finite iff h is integrable

In this sense any measurable non-negative function is a density. In fact, examples of ***signed measures*** may also be constructed in this way if the non-negativity assumption is released.

The measure ν is a generalization of those $\mathcal{LS}$-measures μ_s that have densities; indeed ν is:

Definition 2.3.17. *A measure ν is **absolutely continuous**, with respect to a measure μ (and denoted by $\nu \ll \mu$), whenever*

$$\textit{if} \quad S \in \mathcal{S} \quad \textit{is such that} \quad \mu(S) = 0 \qquad \text{then} \qquad \nu(S) = 0\,.$$

An $\mathcal{LS}$-measure μ_s is absolutely continuous with respect to the Lebesgue measure λ if and only if its generating distribution function s is an absolutely continuous function on $\mathbb{R}^n$; this is the case if and only if $s \equiv s_c$ admits a density function f, see Eq. (2.2.10).

A powerful extension of this property is the following

THEOREM 2.3.7. ***(Radon Nikodym)*** *([ROa] Chapter 11). Let $(\mathbf{X}, \mathcal{S}, \mu)$ be σ-finite and let $\nu \ll \mu$ be a σ-finite measure defined on $\mathcal{S}$. Then there exists a non-negative measurable function $f : (\mathbf{X}, \mathcal{S}, \mu) \to \overline{\mathbb{R}}_+$ such that*

$$\nu(S) = \int_S f \, d\mu, \qquad S \in \mathcal{S}. \tag{2.3.20}$$

Any other function satisfying these properties is μ-a.e. equal to f.

The function f is called ***Radon Nikodym derivative*** or ***density*** of ν with respect to μ (denoted by $d\nu/d\mu$). Moreover: if h is a finite valued measurable function from $(\mathbf{X}, \mathcal{S}, \mu)$ to $\mathbb{R}$ then

$$\int_{\mathbf{X}} h d\nu = \int_{\mathbf{X}} h \frac{d\nu}{d\mu} \, d\mu . \tag{2.3.21}$$

In the case of $\mathcal{LS}$-measures the Radon Nikodym function is the derivative of the (absolutely continuous term of) the generating distribution function.

It is interesting to recall that any σ-finite measure ν on a σ-finite measure space $(\mathbf{X}, \mathcal{S}, \mu)$ may be uniquely decomposed into the sum of two measures: $\nu = \nu_{ac} + \nu_s$, where $\nu_{ac} \ll \mu$ and $\nu_s \perp \mu$, this last formula means that there exists $S \in \mathcal{S}$ such that $\nu_s(S) = \mu(\mathbf{X} \setminus S) = 0$.

Extensions may also be considered to the vector case $f := (f^1, \ldots, f^n)$ where $f^i : (\mathbf{X}^i, \mathcal{S}^i, \mu^i) \to \overline{\mathbb{R}}$ for $i = 1, \ldots, n$. Definitions of simple functions and measurable functions allow immediate componentwise generalizations. For instance:

- ***Random vectors*** $r : (\mathbf{X}, \mathcal{S}, \mu) \to \overline{\mathbb{R}}^n$ are defined componentwise.
- ***Joint marginal measure relative to*** r or ***distribution of*** r is the measure μ_r on $(\overline{\mathbb{R}}^n, \mathcal{B}(\overline{\mathbb{R}}^n))$ given by

$$\mu_r(B) := \mu\left(r^{-1}(B)\right), \qquad B \in \mathcal{B}(\overline{\mathbb{R}}^n). \tag{2.3.22}$$

- ***Generating distribution function of the random vector*** r is the function $s_r : \mathbb{R}^n \to \mathbb{R}_+$ defined by

$$s_r(x) := \mu_r(-\infty, x] := \mu\left(r^{-1}\left((-\infty, x]\right)\right), \quad x \in \mathbb{R}^n.$$

A further way to generate a (particular) measure μ_T on a measurable space $\left(\mathbf{X}^2, \mathcal{S}^2\right)$ starting from a given measure space $\left(\mathbf{X}^1, \mathcal{S}^1, \mu\right)$ resides in the following procedure.

Let $T : \left(\mathbf{X}^1, \mathcal{S}^1, \mu\right) \to \left(\mathbf{X}^2, \mathcal{S}^2\right)$ be a measurable transformation. The following generalizes formula (2.3.22) and defines a measure on $\mathcal{S}^2$:

$$\mu_T(S^2) \;:=\; \mu\left(T^{-1}(S^2)\right), \qquad S^2 \in \mathcal{S}^2. \tag{2.3.23}$$

It is clear that even if $\left(\mathbf{X}^2, \mathcal{S}^2\right)$ were a copy of $\left(\mathbf{X}^1, \mathcal{S}^1\right)$ then μ_T does not necessarily coincide with μ. In this case T is an example of:

Definition 2.3.18. ***Measure preserving transformation*** T *is a measurable transformation* $T : (\mathbf{X}, \mathcal{S}, \mu) \to (\mathbf{X}, \mathcal{S}, \mu)$ *such that*

$$\mu(S) \;=\; \mu\left(T^{-1}(S)\right), \qquad S \in \mathcal{S}.$$

However, even if T does not preserve the measure, the constructions seen above may be composed, for example as it follows, and in each case produces new measures.

Example 2.3.11. Let T be a measurable transformation from $(\mathbf{X}^1, \mathcal{S}^1, \mu)$ to $(\mathbf{X}^2, \mathcal{S}^2)$ and let μ_T be defined as in Eq. (2.3.23). If $h : \mathbf{X}^2 \to \overline{\mathbf{R}}_+$ is measurable, then the following defines a measure on $\mathcal{S}^2$ if either of the two integrals on the right-hand side exists

$$\nu(S^2) = \int_{S^2} h \, d\mu_T = \int_{T^{-1}S^2} h \circ T \, d\mu , \qquad S^2 \in \mathcal{S}^2 .$$

□

Example 2.3.12. Let $(\mathbf{X}^2, \mathcal{S}^2, \mu^2)$ be σ-finite, $g : (\mathbf{X}^2, \mathcal{S}^2) \to \overline{\mathbf{R}}$ a measurable function, and $T : (\mathbf{X}^1, \mathcal{S}^1, \mu^1) \to (\mathbf{X}^2, \mathcal{S}^2, \mu^2)$ a measurable mapping. If the measure $\mu_T := \mu^1 \circ T^{-1}$, defined as in Eq. (2.3.23), happens to be absolutely continuous with respect to μ^2 then its Radon Nikodym derivative

$$\varphi := \frac{d\mu_T}{d\mu^2} : (\mathbf{X}^2, \mathcal{S}^2, \mu^2) \to \overline{\mathbf{R}}$$

satisfies

$$\int_{\mathbf{X}^2} g\varphi \, d\mu^2 = \int_{T^{-1}\mathbf{X}^2} g \circ T \, d\mu^1 .$$

(We remark that the condition $\mu^1 \circ T^{-1} \ll \mu^2$ is essentially a regularity condition on T^{-1}; in real spaces with Lebesgue measure this yields T^{-1} being of bounded variation on compacts and hence a.e. differentiable with an integrable derivative: $\varphi = Jac(T^{-1})$.)

□

It is also of interest to note that what was done to construct $\mathcal{LS}$-measures of $(a, b]$ when $a, b \in \mathbf{R}^n$ is in fact an example of the Classical Product Measure Theorem. To recall it, let us consider a finite collection of measure spaces $(\mathbf{X}^i, \mathcal{S}^i, \mu^i)$, $i = 1, \ldots, n$, and let $\mathbf{X} := \prod_{i=1}^n \mathbf{X}^i$.

Definition 2.3.19. ***Measurable rectangle*** *is a set* $S := \prod_{i=1}^{n} S^i$, *where* $S^i \in \mathcal{S}^i$ *for each* $i = 1, \ldots, n$.

The collection of all measurable rectangles is a semialgebra.

Definition 2.3.20. ***Product*** σ***-algebra*** $\mathcal{S}$ *(improperly denoted by* $\mathcal{S}^1 \times \ldots \times \mathcal{S}^n$*) is the minimal* σ*-algebra in* $2^{\mathbf{X}^1 \times \ldots \times \mathbf{X}^n}$ *generated by the semi-algebra of measurable rectangles.*

The following theorem is here stated in the case $n = 2$ ([ASa] Theorem 2.6.2).

THEOREM 2.3.8. ***(Product Measure)*** *Let* $(\mathbf{X}^1, \mathcal{S}^1, \mu^1)$ *be a* σ*-finite measure space,* $(\mathbf{X}^2, \mathcal{S}^2)$ *be a measure space,* $\mathbf{X} := \mathbf{X}^1 \times \mathbf{X}^2$ *and* $\mathcal{S} := \mathcal{S}^1 \times \mathcal{S}^2$. *If there exists a family* $\{\mu|_{x^1}\}_{x^1 \in \mathbf{X}^1}$ *of* σ*-finite (uniformly with respect to* x^1*) measures on* $\mathcal{S}^2$, *such that the function* $\mu|_{x^1}(S^2) : x^1 \mapsto \overline{\mathbb{R}}_+$ *is* $\mathcal{S}^1$*-measurable for each* $S^2 \in \mathcal{S}^2$, *then the set function defined on* $\mathcal{S}$ *by*

$$\mu(S) := \int_{\mathbf{X}^1} \mu|_{x^1}(S|_{x^1})\, d\mu^1, \quad S \in \mathcal{S}, \tag{2.3.24}$$

where

$$S|_{x^1} := \left\{ x^2 \in \mathbf{X}^2 \,\middle|\, (x^1, x^2) \in S \right\} \in \mathcal{S}^2,$$

is σ*-finite and is the unique measure on* $(\mathbf{X}, \mathcal{S})$ *such that*

$$\mu(S^1 \times S^2) = \int_{S^1} \mu|_{x^1}(S^2)\, d\mu^1, \quad S^i \in \mathcal{S}^i, \quad i = 1, 2.$$

THEOREM 2.3.9. ***(Classical Product Measure)*** *If* $(\mathbf{X}^i, \mathcal{S}^i, \mu^i)$, $i = 1, 2$, *are* σ*-finite measure spaces then the assumption of the preceding theorem are satisfied by the family* $\{\mu|_{x^1}\}_{x^1 \in \mathbf{X}^1} := \mu^2$ *and the classical formula follows, uniquely extending on* $\mathcal{S}$ *the measure*

μ *such that* $\mu(S^1 \times S^2) = \mu^1(S^1) \cdot \mu^2(S^2)$

$$\mu(S) \;=\; \int_{\mathbf{X}^1} \mu^2(S|_{x^1})\, d\mu^1 \;\equiv\; \int_{\mathbf{X}^2} \mu^1(S|_{x^2})\, d\mu^2 \,.$$

This is the reason why the construction seen in Eq. (3.3.24) is sometimes denoted by $(\mu^1 \times \mu^2)(S)$.

THEOREM 2.3.10. ***(Fubini)*** *Let* $(\mathbf{X}^i, \mathcal{S}^i, \mu^i)$, $i = 1,2$, *be two* σ*-finite measure spaces and let* $\mathbf{X} := \mathbf{X}^1 \times \mathbf{X}^2$, $\mathcal{S} := \mathcal{S}^1 \times \mathcal{S}^2$, $\mu := \mu^1 \times \mu^2$ *defined as in Eq. (3.3.24). If* $f : (\mathbf{X}, \mathcal{S}, \mu) \to \overline{\mathbb{R}}$ *is* $\mathcal{S}$*-measurable and integrable, then:*

a) for a.a. $x^1 \in \mathbf{X}^1$ *the functions* $f(x^1, \cdot) : \mathbf{X}^2 \to \overline{\mathbb{R}}$ *are* $\mathcal{S}^2$*-measurable and integrable;*

b) the integrals $f^1(x^1) := \int_{\mathbf{X}^2} f(x^1, \cdot)\, d\mu^2$ *exist a.e. and define a.e. a function* $f^1 : (\mathbf{X}^1, \mathcal{S}^1, \mu^1) \to \overline{\mathbb{R}}$ *that is* $\mathcal{S}^1$*-measurable and integrable;*

c) the same relations (a) and (b) are true by replacing 1 *with* 2*;*

d) the following holds true

$$\int_{\mathbf{X}} f\, d\mu = \int_{\mathbf{X}^1} \left(\int_{\mathbf{X}^2} f(x^1, x^2)\, d\mu^2 \right) d\mu^1$$
$$= \int_{\mathbf{X}^2} \left(\int_{\mathbf{X}^1} f(x^1, x^2)\, d\mu^1 \right) d\mu^2 \,. \qquad (2.3.25)$$

2.3.5 Two conclusive examples

In this subsection we only sketch, and introduce the pertinent notation, of the following two examples. They may be considered, in a sense, conclusive of the theory and represent the main (and ultimate) motivation of this chapter. They are also the reference points of the following chapters.

Example 2.3.13. The model of N isolated identical particles, or ***the canonical ensemble***, [KGa]. The phase space is $\mathbf{X} = \Lambda_\mathbf{x} \times \Lambda_\mathbf{v}$ and the σ-algebra $\mathcal{S} = \mathcal{B}(\mathbf{X})$; sets $\Lambda_\mathbf{x}$ and $\Lambda_\mathbf{v}$ are homeomorphic with subsets of $\mathbb{R}^{2\nu N}$, where ν is the degrees of freedom of the particles, and represent the configuration vessel and the velocities range respectively. In what follows we shall assume, for simplicity, that $\Lambda_\mathbf{x} \equiv \Lambda_\mathbf{v} = \mathbb{R}^{3N}$. A measure μ is defined on $(\mathbf{X}, \mathcal{B}(\mathbf{X}))$ to take account of the occupancy probabilities of the system states. For instance,

$$\int_\mathbf{X} \chi_B \, d\mu \,, \qquad B \in \mathcal{B}(\mathbf{X})$$

represents the probability that the system assumes (any) one of the states $(\mathbf{x}, \mathbf{v})$ in the set B of the phase space $\Lambda_\mathbf{x} \times \Lambda_\mathbf{v}$. Furthermore, in the case when the measure has a density F_N with respect to Lebesgue measure, the product $F_N(\mathbf{x}, \mathbf{v}) d\lambda$ represents the probability that the system is found in any of the states of the elementary volume $d\lambda$ centered at point $(\mathbf{x}, \mathbf{v})$ of the phase space. We remark that

$$(\mathbf{x}, \mathbf{v}) \equiv (\mathbf{x}^1, \ldots, \mathbf{x}^N; \quad \mathbf{v}^1, \ldots, \mathbf{v}^N)$$

represents the microscopical state of the system, $\mathbf{x} \in \mathbb{R}^{3N}$ the configuration vector, $\mathbf{v} \in \mathbb{R}^{3N}$ the velocity vector, each $(\mathbf{x}^i, \mathbf{v}^i) \in \mathbb{R}^6$, for $i = 1, \ldots, N$, the state vectors of the i-th particle, and hence

$$F_N(t, \mathbf{x}^1, \ldots, \mathbf{x}^N, \mathbf{v}^1, \ldots, \mathbf{v}^N) \, d\mathbf{x}^1 \cdots d\mathbf{x}^N \, d\mathbf{v}^1 \cdots d\mathbf{v}^N$$

denotes the probability that the i-th particle is found in any of the states of the set $d\mathbf{x}^i \, d\mathbf{v}^i$ centered at $(\mathbf{x}^i, \mathbf{v}^i)$, for each $i = 1, \ldots, N$. The further hypothesis that the system consists of indistinguishable particles is modelled by assuming that the function $F_N(\mathbf{x}^1, \ldots, \mathbf{x}^N; \mathbf{v}^1, \ldots, \mathbf{v}^N)$ is symmetric with respect to all the $N!$ permutations:

$$(\mathbf{x}^1, \ldots, \mathbf{x}^N) \mapsto (\mathbf{x}^{i_1}, \ldots, \mathbf{x}^{i_N}) =: (\mathbf{x}_1, \ldots, \mathbf{x}_N) \,,$$

together with

$$(\mathbf{v}^1, \ldots, \mathbf{v}^N) \mapsto (\mathbf{v}^{i_1}, \ldots, \mathbf{v}^{i_N}) =: (\mathbf{v}_1, \ldots, \mathbf{v}_N),$$

and that to a single cloud of N particles in the $\mathbb{R}^6$-space there correspond $N!$ points in the $\mathbb{R}^{6N}$-space. Hence, $(\mathbf{x}_1, \ldots, \mathbf{x}_N, \mathbf{v}_1, \ldots, \mathbf{v}_N)$ denotes the state vector of such a cloud and, by defining

$$f_N(t, \mathbf{x}_1, \ldots, \mathbf{x}_N, \mathbf{v}_1, \ldots, \mathbf{v}_N) := N!\, F_N(t, \mathbf{x}^1, \ldots, \mathbf{x}^N, \mathbf{v}^1, \ldots, \mathbf{v}^N), \tag{2.3.26}$$

one has

$$f_N(t, \mathbf{x}_1, \ldots, \mathbf{x}_N, \mathbf{v}_1, \ldots, \mathbf{v}_N)\, d\mathbf{x}_1 \cdots d\mathbf{x}_N\, d\mathbf{v}_1 \ldots d\mathbf{v}_N$$

denotes the expected number of ways in which, at time t, the N-particle system may be found with a (whatever) molecule in any of the states of the set $d\mathbf{x}_i\, d\mathbf{v}_i$ centered at $(\mathbf{x}_i, \mathbf{v}_i) \in \mathbb{R}^6$, for each $i = 1, \ldots, N$. In formula (2.3.26) F_N is computed in (only) one of its $N!$ possible permutations, and the symmetry of F_N allows to ignore which one. The definition implies

$$\int_{\mathbb{R}^{6N}} f_N(t, \mathbf{x}_1, \ldots, \mathbf{x}_N, \mathbf{v}_1, \ldots, \mathbf{v}_N)\, d\mathbf{x}_1 \cdots d\mathbf{x}_N\, d\mathbf{v}_1 \ldots d\mathbf{v}_N = N!\,.$$

With the same notation, the following may be introduced

$$\begin{aligned} & f_{N,k}(t, \mathbf{x}_1, \ldots, \mathbf{x}_k, \mathbf{v}_1, \ldots, \mathbf{v}_k) \\ & \quad = \frac{N!}{(N-k)!} \int_{\mathbb{R}^{6(N-k)}} F_N(t, \mathbf{x}^1, \ldots, \mathbf{x}^N, \mathbf{v}^1, \ldots, \mathbf{v}^N) \\ & \quad \times d\mathbf{x}^{k+1} \cdots d\mathbf{x}^N\, d\mathbf{v}^{k+1} \cdots d\mathbf{v}^N \end{aligned} \tag{2.3.27}$$

for $k = 1, \ldots, N-1$. Hence, one has

$$\begin{aligned} & f_{N,k}(t, \mathbf{x}_1, \ldots, \mathbf{x}_k, \mathbf{v}_1, \ldots, \mathbf{v}_k) \\ & \quad = \frac{N!}{(N-k)!}\, F_k(t, \mathbf{x}^1, \ldots, \mathbf{x}^k, \mathbf{v}^1, \ldots, \mathbf{v}^k) \\ & \quad = \frac{N!}{k!\,(N-k)!}\, f_k(t, \mathbf{x}_1, \ldots, \mathbf{x}_k, \mathbf{v}_1, \ldots, \mathbf{v}_k) \\ & \quad = \frac{1}{(N-k)!} \int_{\mathbb{R}^{6(N-k)}} f_N(t, \mathbf{x}_1, \ldots, \mathbf{x}_N, \mathbf{v}_1, \ldots, \mathbf{v}_N) \\ & \quad \times d\mathbf{x}_{k+1} \cdots d\mathbf{x}_N\, d\mathbf{v}_{k+1} \cdots d\mathbf{v}_N \end{aligned}$$

where the usual normalization conditions on the k-marginal F_k of F_N have been maintained for all $k = 1, 2, \ldots, N$:

$$\int_{\mathbb{R}^{6k}} F_k(t, \mathbf{x}^1, \ldots, \mathbf{x}^k, \mathbf{v}^1, \ldots, \mathbf{v}^k)\, d\mathbf{x}^1 \cdots d\mathbf{x}^k\, d\mathbf{v}^1 \cdots d\mathbf{v}^k = 1\,, \tag{2.3.28}$$

which imply, for $1 \le j \le k \le N$ and $f_{N,0} := 1$,

$$\begin{aligned} & \int_{\mathbb{R}^{6j}} f_{N,k}(t, \mathbf{x}_1, \ldots, \mathbf{x}_k, \mathbf{v}_1, \ldots, \mathbf{v}_k)\, d\mathbf{x}_{k-j+1} \cdots d\mathbf{x}_k\, d\mathbf{v}_{k-j+1} \ldots d\mathbf{v}_k \\ & = \frac{(N-k+j)!}{(N-k)!}\, f_{N,k-j}(t, \mathbf{x}_1, \ldots, \mathbf{x}_{k-j}, \mathbf{v}_1, \ldots, \mathbf{v}_{k-j})\,. \end{aligned} \tag{2.3.29}$$

Definition 2.3.21. ***k-particle (statistical) distribution*** *for the N-particle system is the function $f_{N,k}$ defined by Eq. (2.2.27). It denotes the number density of the possible ways in which k out of the N particles may be found with a particle in the state $(\mathbf{x}_i, \mathbf{v}_i)$ for each $i = 1, \ldots, k$.*

Equivalently, the product

$$f_{N,k}(t, \mathbf{x}_1, \ldots, \mathbf{x}_k, \mathbf{v}_1, \ldots, \mathbf{v}_k)\, d\mathbf{x}_1 \cdots d\mathbf{x}_k\, d\mathbf{v}_1 \ldots d\mathbf{v}_k$$

gives the expected number of k-tuples that the N-particle system shows to have with states in $d\mathbf{x}_i\, d\mathbf{v}_i$ centered at $(\mathbf{x}_i, \mathbf{v}_i)$, for $i = 1, \ldots, k$. In particular one has:

Definition 2.3.22. ***i-th particle density function*** *is the 1-marginal density function* $F_1 : \mathbb{R} \times \mathbb{R}^6 \to \mathbb{R}_+$:

$$F_1(t, \mathbf{x}^i, \mathbf{v}^i) = \int_{\mathbb{R}^{6(N-1)}} F_N(t, \mathbf{x}^1, \ldots, \mathbf{x}^N, \mathbf{v}^1, \ldots, \mathbf{v}^N) d\mathbf{x}^2 \cdots d\mathbf{x}^{i-1} \times d\mathbf{x}^{i+1} \cdots d\mathbf{x}^N d\mathbf{v}^2 \cdots d\mathbf{v}^{i-1} d\mathbf{v}^{i+1} \cdots d\mathbf{v}^N .$$

The value $F_1(t, \mathbf{x}^i, \mathbf{v}^i) d\mathbf{x}^i d\mathbf{v}^i$ represents the probability that the i-th one of the N particles has its state in the elementary cell $d\mathbf{x}^i\, d\mathbf{v}^i$ centered at $(\mathbf{x}^i, \mathbf{v}^i)$. With the above notation one has

$$F_1(t, \mathbf{x}^1, \mathbf{v}^1) d\mathbf{x}^1\, d\mathbf{v}^1 = F_1(t, \mathbf{x}^i, \mathbf{v}^i) d\mathbf{x}^i\, d\mathbf{v}^i = f_1(t, \mathbf{x}_1, \mathbf{v}_1) d\mathbf{x}_1\, d\mathbf{v}_1 ,$$

where f_1 is the one-particle system probability density function.

Definition 2.3.23. *One-particle (statistical)* ***distribution function,*** *or* ***number density*** *for an N-particle system is the function $f_{N,1} : \mathbb{R} \times \mathbb{R}^6 \to [0, N]$ given by formula (2.3.27) where $k = 1$, namely:*

$$f_{N,1}(t, \mathbf{x}_1, \mathbf{v}_1) := N \int_{\mathbb{R}^{6(N-1)}} F_N(t, \mathbf{x}, \mathbf{v})\, d\mathbf{x}^2 \cdots d\mathbf{x}^N\, d\mathbf{v}^2 \cdots d\mathbf{v}^N . \tag{2.3.30}$$

Hence: $\quad f_{N,1}(t, \mathbf{x}_1, \mathbf{v}_1) = N\, f_1(t, \mathbf{x}_1, \mathbf{v}_1)$, $\quad$ and

$$f_{N,1}(t, \mathbf{x}_1, \mathbf{v}_1) = \frac{1}{N-1} \int_{\mathbb{R}^6} f_{N,2}(t, \mathbf{x}_1, \mathbf{v}_1, \mathbf{x}_2, \mathbf{v}_2)\, d\mathbf{x}_2\, d\mathbf{v}_2 .$$

The product $f_{N,1}(t, \mathbf{x}_1, \mathbf{v}_1) d\mathbf{x}_1\, d\mathbf{v}_1$ represents the ***expected number*** (see Eq. (2.3.33)) of particles that have states in the elementary cell $d\mathbf{x}_1\, d\mathbf{v}_1$ centered at $(\mathbf{x}_1, \mathbf{v}_1)$.

The function $f_{N,1}$ is often denoted by f without reference to the (unknown although constant and finite) total number of particles N.

□

Example 2.3.14. The model of a non-isolated $N = N(t)$-particle system in classical statistical mechanics, or: the ***grand canonical ensemble***, [RUa]. In the mechanical scheme, a commonly accepted probability density function, or ***thermodynamic state***, is modelled by the ***Gibbs measure*** γ. It represents a system at the thermodynamic equilibrium with a heat reservoir. The measure γ is given by

$$d\gamma := \frac{1}{N!}\,\frac{1}{Z_\Lambda}\, e^{\beta N u}\, e^{-\beta \mathcal{H}(p,q)} \chi_\Lambda(q)\; dp\, dq\,;$$

where the notation is: $N \in \mathbb{N}$ is the (variable) number of particles, (p,q) is the state vector in $\mathbf{X} := \mathbb{R}^{\nu N} \times \Lambda \subseteq \mathbb{R}^{2\nu N}$, ν the number of degrees of freedom, $\mathbf{X}$ is symmetrized by permutations and endowed with the Lebesgue measure $dp\,dq := d\lambda$ on $\mathcal{B}(\mathbf{X})$; $\mathcal{H}$ is the Hamiltonian function on $\mathbf{X}$; $\beta = (k\tau)^{-1}$ where k is the Boltzmann constant and τ the absolute temperature of the reservoir; u is the chemical potential and Z_Λ the so-called ***partition function***, i.e., the normalization condition:

$$Z_\Lambda := 1 + \sum_{N=1}^{\infty} \int_{\mathbf{X}} \frac{1}{N!}\, e^{\beta N u}\, e^{-\beta \mathcal{H}(p,q)} \chi_\Lambda(q)\; dp\, dq\,.$$

□

The following definitions may finally be given.

Definition 2.3.24. ***Physical observable*** g *is an integrable function* $g : (\mathbf{X}, \mathcal{S}, \mu) \to \mathbb{R}$ *defined on a* σ*-finite measure space* $(\mathbf{X}, \mathcal{S}, \mu)$ *locally homeomorphic with* $(\mathbb{R}^m, \mathcal{B}(\mathbb{R}^m), \lambda)$.

In particular, any integrable real valued random vector $r : \mathbf{X} \to \mathbb{R}^n$ defined on such a probability space may represent a physical observable. Similarly, any real valued Borel-measurable function $g : \mathbf{R}^n \to \mathbf{R}$ of a random vector $r : \mathbf{X} \to \mathbf{R}^n$ is again a random variable, and as such it may represent a physical observable.

Definition 2.3.25. ***Mean value*** *$E[g \circ r]$, or* ***observed value****, or* ***expectation*** *of a physical quantity represented by a function $g : \mathbf{R} \to \mathbf{R}$ of a random variable $r : x \in \mathbf{X} \mapsto u = r(x) \in \mathbf{R}$, is the value shared by the $\mathcal{LS}$-integrals*

$$E[g \circ r] := \int_{\mathbf{X}} g \circ r \, d\mu = \int_{\mathbf{R}} g \, d\mu_r \equiv \int_{\mathbf{R}} g(u) \, ds_r(u) \qquad (2.3.31)$$

at the (equilibrium) thermodynamic state μ.

Actually, see Definition 2.2.22 and Example 2.3.12, (or [ASa] Theorem 5.10.2), whenever the first of these integrals exists so does the other and they are equal.

Definition 2.3.26. ***Moments of order*** *$k \in \mathbf{N}$ of a real valued random variable $r : \mathbf{X} \to \mathbf{R}$ is the expectation of the random variable*

$$g : r \to r^k .$$

In the following chapters, the measure μ_r will in general admit a density $f : u = r(x) \in \mathbf{R} \mapsto \mathbf{R}_+$ with respect to the Lebesgue measure, and the density itself will be the unknown of the problem. In this case, one has $d\mu_r = f(u)\, du$, and the expectation of a function $g \circ r$ will be denoted by

$$E_g[f] = \int_{\mathbf{R}} g(u)\, f(u)\, du . \qquad (2.3.32)$$

In particular, the moments of order k of the variable r will be denoted by $E_k[f]$. For instance, if $f = f(t, u)$ is a (one-parameter family of) statistical distribution functions with respect to the random variable

u, then one has

$$E_0[f](t,u) \equiv N(t) = \int_{\mathbb{R}} f(u)\, du \tag{2.3.33}$$

i.e., the total number of individuals of the system. Similarly, the mean value of r will be

$$E[r](t) = \frac{E_1[f](t)}{E_0[f](t)} . \tag{2.3.34}$$

2.4 Tools from Topology

In this section we provide the background for some convergence methods such as those applied to the qualitative analysis that has been developed in Chapter 4 with reference to coagulation fragmentation models. Topological methods have recently been used also for the qualitative analysis of the Cauchy problem for strict Boltzmann models. The reader interested in these methodologies, and in the analytical aspects of the nonlinear kinetic models evolution, may take advantage of the following, however concise, presentation.

2.4.1 Basics on topologies

We first recall some introductory definitions, see [KEa].

Let $\mathbf{X}$ be a set, $2^{\mathbf{X}}$ the family of subsets of $\mathbf{X}$, and O the ***open sets*** of $\mathbf{X}$, i.e., the elements of the ***topology*** $\mathcal{T} \subset 2^{\mathbf{X}}$, which is a family of subsets of $\mathbf{X}$ containing $\mathbf{X}, \emptyset$, and closed under unions and finite intersections. $(\mathbf{X}, \mathcal{T})$ is said to be a ***topological space***.

Definition 2.4.1. ***Neighborhood*** *$\mathcal{N}(A)$ of a set $A \in 2^{\mathbf{X}}$, is a set $\mathcal{N}(A) \in 2^{\mathbf{X}}$ such that $A \subset O \subset \mathcal{N}(A)$ for some $O \in \mathcal{T}$.*

The space $(\mathbf{X}, \mathcal{T})$ is said

T1-*space* when for any pair $x, y \in \mathbf{X}$, $x \neq y$, there exist neighborhoods $\mathcal{N}(x) \not\ni y$ and $\mathcal{N}(y) \not\ni x$. This condition is equivalent to require that each singleton $\{x\}$ is a closed set.

T2-*space* or ***Hausdorff*** when for any pair $x, y \in \mathbf{X}$, $x \neq y$, there exist disjoint neighborhoods $\mathcal{N}(x)$ and $\mathcal{N}(y)$. This condition is equivalent to uniqueness of the limit points of nets (see later).

T3-*space* when it is **T1** and ***regular***, i.e., for any $x \in \mathbf{X}$, and $K \in 2^{\mathbf{X}}$ closed, $K \not\ni x$, there exist disjoint neighborhoods $\mathcal{N}(x)$ and $\mathcal{N}(K)$.

T$\frac{3}{2}$-*space* when it is **T1** and ***completely regular***, i.e., for any $x \in \mathbf{X}$ and $\mathcal{N}(x)$ open, there exists a continuous $f : \mathbf{X} \to [0,1]$ such that $f(x) = 1$ and $f(\mathbf{X} \setminus \mathcal{N}(x)) = 0$.

T4-*space* when it is **T1** and ***normal***, i.e., for any pair of closed, disjoint, sets $K_1, K_2 \in 2^{\mathbf{X}}$ there exist disjoint neighborhoods $\mathcal{N}(K_1)$ and $\mathcal{N}(K_2)$.

Definition 2.4.2. ***Base*** *for the topology $\mathcal{T}$ is a family $\mathcal{B}_{\mathcal{T}}$ of open sets such that for any $x \in \mathbf{X}$ and $O \in \mathcal{T}$, $O \ni x$, there exists $B \in \mathcal{B}_{\mathcal{T}}$ such that $x \in B \subseteq O$. Equivalently: each $O \in \mathcal{T}$ is the union of sets in $\mathcal{B}_{\mathcal{T}}$.*

Definition 2.4.3. ***Subbase*** *for the topology $\mathcal{T}$ is a family $\mathcal{B}_{\mathcal{B}}$ of open sets such that the family of finite intersections of elements of $\mathcal{B}_{\mathcal{B}}$ is a base for $\mathcal{T}$.*

Let $\{\mathbf{X}_\alpha, \mathcal{T}_{\mathbf{X}_\alpha}\}_{\alpha \in A}$ be a family of topological spaces, and denote by $\mathbf{Y}$ the product space $\mathbf{Y} := \bigtimes_{\alpha \in A} \mathbf{X}_\alpha$.

Definition 2.4.4. ***Product topology*** *$\mathcal{T}^{\Pi}$ is the smallest topology on $\mathbf{Y}$ such that the projections $\pi_\alpha : y \mapsto y_\alpha$ are continuous for each $\alpha \in A$.*

A subbase for $\mathcal{T}^{\Pi}$ is the family of sets of the form $\pi_\alpha^{-1}[O_\alpha]$, for $O_\alpha \in \mathcal{T}_{\mathbf{X}_\alpha}$ and $\alpha \in A$.

Definition 2.4.5. ***Partial order*** *on a set* $\mathbf{X}$ *is a reflexive and transitive relation: "* $\leq$ *" on the points of the set* $\mathbf{X}$. *(The order is total when all the pairs of elements in* $\mathbf{X}$ *are related by the order relation.)*

Definition 2.4.6. ***Directed set*** D *is a partially ordered set such that, whenever* a, b *are in* D *then there exists* $c \in D$ *with* $a \leq c$ *and* $b \leq c$.

Definition 2.4.7. ***Net***, *or* ***generalized sequence***, *is a function on a directed set.*

Definition 2.4.8. ***Uniformity*** $\mathcal{U} \subset 2^{\mathbf{X}\times\mathbf{X}}$ *is a family of sets such that*

a) $U \in \mathcal{U} \quad \Rightarrow \quad U \ni \Delta := \{(x,x) \mid x \in \mathbf{X}\}$;

b) $U \in \mathcal{U} \quad \Rightarrow \quad U^{-1} \in \mathcal{U}$, where $U^{-1} := \{(y,x) \mid (x,y) \in U\}$;

c) $U \in \mathcal{U} \quad \Rightarrow \quad$ *there exists* $V \in \mathcal{U}$ *such that* $V \circ V \subset U$, *where:*

$$V_1 \circ V_2 := \{(x,z) \,\big|\, \exists y : \quad (x,y) \in V_1, \quad (y,z) \in V_2\} ;$$

d) $U, V \in \mathcal{U} \quad \Rightarrow \quad U \cap V \in \mathcal{U}$;

e) $U \in \mathcal{U}$, and $U \subset V \subset \mathbf{X} \times \mathbf{X}, \quad \Rightarrow \quad V \in \mathcal{U}$.

Definition 2.4.9. ***Base*** *for the uniformity* $\mathcal{U}$ *is a family* $\mathcal{B}_{\mathcal{U}}$ *of sets such that each* $U \in \mathcal{U}$ *contains a set* $B \in \mathcal{B}_{\mathcal{U}}$.

Definition 2.4.10. ***Subbase*** *for the uniformity* $\mathcal{U}$ *is a family* $\mathcal{B}^*_{\mathcal{U}}$ *of sets such that the family of finite intersections of set in* $\mathcal{B}^*_{\mathcal{U}}$ *is a base* $B \in \mathcal{B}_{\mathcal{U}}$ *for the uniformity* $\mathcal{U}$.

Definition 2.4.11. ***(Uniform) topology*** $\mathcal{T}_{\mathcal{U}}$, *or* ***topology of the uniformity*** $\mathcal{U}$, *is the topology defined by*

$$\mathcal{T}_{\mathcal{U}} := \left\{ O \in 2^{\mathbf{X}} \,\middle|\, x \in O \Rightarrow \exists U \in \mathcal{U} \quad \text{such that} \quad U[x] \subset O \right\}, \tag{2.4.1}$$

where

$$U[x] := \{x' \in \mathbf{X} \mid (x', x) \in U\}. \tag{2.4.2}$$

Example 2.4.1. Metric uniformity is the subset of $2^{\mathbf{X}\times\mathbf{X}}$ which contains the sets

$$V_{d,r} := \left\{(x, x') \in \mathbf{X} \times \mathbf{X} \;\middle|\; d(x, x') < r\right\} \qquad \text{for} \quad r \in \mathbb{R}_+. \tag{2.4.3}$$

Metric topology is the family of all sets which contain a sphere about each point.

□

A topological space $(\mathbf{X}, \mathcal{T})$ is completely regular if and only if $\mathcal{T}$ is the uniform topology for some uniformity $\mathcal{U}$ on $\mathbf{X}$. A space $(\mathbf{X}, \mathcal{T}_{\mathcal{U}})$ is Hausdorff if and only if it is $\mathbf{T1}$; or, if and only if $\bigcap\{U \mid U \in \mathcal{U}\} = \Delta$.

Definition 2.4.12. *A function $f : (\mathbf{X},\mathcal{U}) \to (\mathbf{Y}, \mathcal{V})$ is said to be* ***uniformly continuous*** *relative to $\mathcal{U}$ when for each $V \in \mathcal{V}$ the set*

$$f_2^{-1}[V] := \{(x, x') \mid (f(x), f(x')) \in V\}$$

belongs to $\mathcal{U}$, and where $f_2 : \mathbf{X}^2 \to \mathbf{Y}^2$ is defined by

$$f_2(x, x') := (f(x), f(x')) . \tag{2.4.4}$$

Equivalently: a function is uniformly continuous whenever for each $V \in \mathcal{V}$ there exists $U \in \mathcal{U}$ such that $f_2[U] \subset V$.

Example 2.4.2. If $\mathbf{Y} \equiv \mathbb{R}$ and $\mathcal{V}$ is, as usual, the family of all sets $V \subset \mathbb{R}^2$ which contain $\{(x, x') \mid |x - x'| < r\}$ for some $r > 0$, then f is uniformly continuous when for each $r > 0$ there is $U \in \mathcal{U}$ such that $|f(x) - f(x')| < r$ whenever $(x, x') \in U$.

□

If f is uniformly continuous then f is $\mathcal{T}_{\mathcal{U}}$-continuous; namely: $f^{-1}[O_{\mathbf{Y}}] \in \mathcal{T}_{\mathcal{U}}$, for any $O_{\mathbf{Y}} \in \mathcal{T}_{\mathcal{V}}$.

Definition 2.4.13. ***Pseudo-metric*** *d on $\mathbf{X}$ is a symmetric, positive semidefinite, triangular function on $\mathbf{X} \times \mathbf{X}$.*

Let $\{ (\mathbf{X}_\alpha, \mathcal{U}_\alpha) \}_{\alpha \in A}$ be a family of uniform spaces.

Definition 2.4.14. ***Product uniformity*** *$\displaystyle\bigtimes_{\alpha\in A} \mathcal{U}_\alpha$ is the smallest uniformity on the space $\mathbf{Y} := \displaystyle\bigtimes_{\alpha\in A} \mathbf{X}_\alpha$ such that each of the projections $\pi_\alpha : \mathbf{Y} \to \mathbf{X}_\alpha$ is uniformly continuous, for $\alpha \in A$.*

The family $\mathcal{B}^*$ of subsets of $\mathbf{Y}$ of the form

$$\left\{ (x, x') \,\middle|\, (x_\alpha, x'_\alpha) \in U_\alpha, \qquad \alpha \in A, \quad U_\alpha \in \mathcal{U}_\alpha \right\}$$

is a subbase for the product uniformity.

The topology of the product uniformity is the product topology. A function $f : (\mathbf{X}, \mathcal{U}) \to \left(\displaystyle\bigtimes_{\alpha\in A} \mathbf{X}_\alpha, \bigtimes_{\alpha\in A} \mathcal{U}_\alpha \right)$ is uniformly continuous if and only if $\pi \circ f$ is uniformly continuous for each $\alpha \in A$. A pseudo-metric d on $\mathbf{X} \times \mathbf{X}$ is uniformly continuous relative to the product uniformity $\mathcal{U} \times \mathcal{U}$ if and only if the sets $V_{d,r}$ defined in (2.4.3) belong to $\mathcal{U}$ for each $r > 0$.

Conversely, each pseudo-metric ***generates*** a uniformity: the family of sets $V_{d,r}$ may be used as a subbase.

Definition 2.4.15. *$(\mathbf{X}, \mathcal{U})$ is **pseudo-metrizable** when there exists a pseudo-metric d such that $\mathcal{U}$ is generated by d. This is the case if $\mathcal{U}$ has a countable base.*

Definition 2.4.16. *$(\mathbf{X}, \mathcal{U})$ is **metrizable** when $(\mathbf{X}, \mathcal{T}_{\mathcal{U}})$ is* **T1** *and $\mathcal{U}$ has a countable base.*

The topology $\mathcal{T}_{\mathcal{U}}$ of the uniformity generated by a pseudo-metric is identical to the pseudo-metric topology, i.e., the topology that has as a base the family of spheres $V_{d,r}[x]$. Recall ([KEa] Section IV.10) that each pseudo-metric space is normal.

A whole family P of pseudo-metrics may generate a uniformity $\mathcal{U}_P$ on $\mathbf{X}$; this may be done by using the subbase $V_{d,r}$, $d \in P$. $\mathcal{U}_P$ is the smallest uniformity such that all $d \in P$ are uniformly continuous on $\mathbf{X} \times \mathbf{X}$, relative to $\mathcal{U} \times \mathcal{U}$.

In fact ([KEa] Section 6.15) each uniformity for $\mathbf{X}$ is generated by the family of all pseudo-metrics that are uniformly continuous on $\mathbf{X} \times \mathbf{X}$.

Definition 2.4.17. ***Gage*** *G for a uniformity $\mathcal{U}$ is the set of all pseudo-metrics which are uniformly continuous on $\mathbf{X} \times \mathbf{X}$, relative to $\mathcal{U} \times \mathcal{U}$. Clearly one has $\mathcal{U} \equiv \mathcal{U}_G$.*

A family P of pseudo-metrics on $\mathbf{X}$ ***generates*** a gage G in that a pseudo-metric g belongs to G whenever for each $s > 0$ there is $r > 0$ and a finite family $d_1, \ldots, d_n \in P$ such that $V_{g,s} \supset \bigcap_{i=1}^{n} V_{d_i,r}$.

Definition 2.4.18. ***Gage Space*** *$(\mathbf{X}, \mathcal{T})$ is a space $\mathbf{X}$ equipped with a topology $\mathcal{T}$ identical to the topology $\mathcal{T}_{\mathcal{U}_P}$ generated by using the family of the sets $V_{d,r}[x]$, for $d \in P$ and $r > 0$, as a subbase.*

Let $G_{\mathcal{U}}$ and $G_{\mathcal{V}}$ be gages for $(\mathbf{X}, \mathcal{U})$ and $(\mathbf{Y}, \mathcal{V})$ respectively. Then $f : (\mathbf{X}, \mathcal{U}) \to (\mathbf{Y}, \mathcal{V})$ is uniformly continuous if and only if for each $d_{\mathcal{V}} \in G_{\mathcal{V}}$ it happens that $d_{\mathcal{V}} \circ f_2$ is a pseudo-metric in $G_{\mathcal{U}}$, where f_2 has been defined in Eq. (2.4.4).
Equivalently: when for each $d_{\mathcal{V}} \in G_{\mathcal{V}}$ and $s > 0$ there exists $d_{\mathcal{U}} \in G_{\mathcal{U}}$ and $r > 0$ such that

$$d_{\mathcal{U}}(x, y) < r \qquad \Rightarrow \qquad d_{\mathcal{V}}\,(f(x), f(y)) < s. \tag{2.4.5}$$

Let $\{(\mathbf{X}_\alpha, \mathcal{U}_\alpha)\}_{\alpha \in A}$, be a set of uniform spaces, and $\{G_\alpha\}_{\alpha \in A}$ the corresponding gages. Then the gage of $\bigtimes_{\alpha \in A} \mathbf{X}_\alpha$ is generated by the family $\{d_\alpha(x_\alpha, y_\alpha) \mid d_\alpha \in G_\alpha, \quad \alpha \in A\}$.

Definition 2.4.19. *A net* $\{x_\beta\}_{\beta \in D}$, $x_\beta \in (\mathbf{X}, \mathcal{U})$, *is said to* ***converge*** *to* $x \in \mathbf{X}$ *when it converges relative to the topology* $\mathcal{T}_\mathcal{U}$*; and hence if and only if*

$$\lim_\beta d_\alpha(x_\beta, x) = 0 \quad \text{for each } d_\alpha \in P,$$

where P *is any subset of the gage* G *that generates* $\mathcal{U}$.

Definition 2.4.20. *A net* $\{x_\beta\}_{\beta \in D}$, $x_\beta \in (\mathbf{X}, \mathcal{U})$, *is said to be a* ***Cauchy net*** *when, for each* $U \in \mathcal{U}$ *there exists* $\beta_0 \in D$ *such that if* $\beta_1, \beta_2 \geq \beta_0$ *then* $(x_{\beta_1}, x_{\beta_2}) \in C$.

Each net that converges, relative to $\mathcal{T}_\mathcal{U}$, to a point $x \in \mathbf{X}$ is a ***Cauchy net***; in particular it is such that for each $r > 0$ it happens that $(x_{\beta_1}, x_{\beta_2})$ belongs to $V_{d,r}$ for sufficiently large β_1, β_2.

2.4.2 Function spaces

Let $\mathbf{F}$ be a family of functions from a set $\mathbf{X}$ to a topological space $(\mathbf{Y}, \mathcal{T}_\mathbf{Y})$. The family $\mathbf{F}$ may be considered as a subset of

$$\mathbf{Y}^\mathbf{X} := \bigtimes_{x \in \mathbf{X}} \mathbf{Y}, \tag{2.4.6}$$

in that there exists a natural correspondence: the ***evaluation*** $\pi : \mathbf{F} \to \mathbf{Y}^\mathbf{X}$ that maps $f \in \mathbf{F}$ into the point $\pi(f) \in \mathbf{Y}^\mathbf{X}$ whose x-th coordinate is $\pi_x(f) := f(x)$.

Definition 2.4.21. ***Relativized projection****, or:* ***evaluation at*** x, *is the* x*-coordinate function* $\pi_x : f \in \mathbf{F} \mapsto f(x) \in \mathbf{Y}$.

π is $1-1$ if and only if $\mathbf{X}$ ***separates the points*** of $\mathbf{F}$, i.e., if and only if for each pair $f, g \in \mathbf{F}$, $f \neq g$, there exists $x \in \mathbf{X}$ such that $f(x) \neq g(x)$. When $\mathbf{F}$ is, for instance, a subset of a functional space $L_1(\mathbf{X}, \mathcal{S}, \mu)$, then the projections π_x may be identified with the (continuous linear) delta-functionals $\widehat{\delta}_x$.

Definition 2.4.22. ***Relativized product topology*** $\mathcal{T}_{\mathbf{F}}^{\Pi}$ *is the smallest topology on* $\mathbf{F} \subset \mathbf{Y}^{\mathbf{X}}$ *such that* π_x *is continuous for each* $x \in \mathbf{X}$.

Relativized in the sense that $\mathbf{F}$ is a subset of $\mathbf{Y}^{\Pi}$, and hence $\mathcal{T}_{\mathbf{F}}^{\Pi} = \mathcal{T}^{\Pi} \cap \mathbf{F}$.

$\mathcal{T}_{\mathbf{F}}^{\Pi}$ coincides with the topology $\mathcal{T}_{\mathbf{F}}^{\mathbf{X}}$ generated (see Definition 2.4.34) on the space $\mathbf{F}$ by the family $\{\pi_x \mid x \in \mathbf{X}\} \longleftrightarrow \mathbf{X}$ of the (coordinate) functions $x : f \in \mathbf{F} \subseteq \mathbf{Y}^{\mathbf{X}} \mapsto f(x)$.

$\mathcal{T}_{\mathbf{F}}^{\Pi} \equiv \mathcal{T}_{\mathbf{F}}^{\mathbf{X}}$ has as a base the family of sets

$$\left\{ f \in \mathbf{F} \,\middle|\, f(x_1) \in O_1, \ldots, f(x_k) \in O_k \right\}$$

for arbitrary $k \in \mathbb{N}$, $x_1, \ldots, x_k \in \mathbf{X}$, and $O_1, \ldots, O_k \in \mathcal{T}_{\mathbf{Y}}$. These definitions are independent of the topology on $\mathbf{X}$, if any.

If $\mathbf{Y}$ is Hausdorff, then so it becomes $(\mathbf{F}, \mathcal{T}_{\mathbf{F}}^{\Pi})$. On the other hand, the product of completely regular **T1** spaces is a completely regular **T1** space.

Definition 2.4.23. *A net* $\{f_\alpha\}_{\alpha \in D}$, $f_\alpha \in \mathbf{F}$, *is said to* ***converge*** *to* $f \in \mathbf{F}$ *with respect to* $\mathcal{T}_{\mathbf{F}}^{\mathbf{X}}$ *when the nets* $\{f_\alpha(x)\}_{\alpha \in D}$ *converge to* $f(x)$ *for each* $x \in \mathbf{X}$. *The topology* $\mathcal{T}_{\mathbf{F}}^{\mathbf{X}}$ *is also called* ***topology of pointwise convergence*** *for* $\mathbf{F} \subseteq \mathbf{Y}^{\mathbf{X}}$.

The following analogous properties are noteworthy (see e.g., [KEa] Section 4.5, [DUa] Section 5.3.2).

Let $\mathbf{F}$ be a family of functions from a set $\mathbf{X}$ to a topological space $(\mathbf{Y}, \mathcal{T}_{\mathbf{Y}})$. The set $\mathbf{X}$ may be considered as a subset of

$$\mathbf{Y}^{\mathbf{F}} := \bigtimes_{f \in \mathbf{F}} \mathbf{Y}, \tag{2.4.7}$$

in that there exists a natural correspondence: the ***evaluation*** $\pi : \mathbf{X} \to \mathbf{Y}^{\mathbf{F}}$ that maps $x \in \mathbf{X}$ into the point $\pi(x) \in \mathbf{Y}^{\mathbf{X}}$ whose f-th coordinate is $\pi_f(x) := f(x)$.

Definition 2.4.24. ***Relativized projection***, *or:* ***evaluation at*** f, *is the f-coordinate function* $\pi_f : x \in \mathbf{X} \mapsto f(x) \in \mathbf{Y}$.

π is 1-1 if and only if $\mathbf{F}$ separates the points of $\mathbf{X}$, i.e., if and only if for each pair $x, x' \in \mathbf{X}$, $x \neq x'$, there exists $f \in \mathbf{F}$ such that $f(x) \neq f(x')$.

Definition 2.4.25. ***Relativized product topology*** $\mathcal{T}_{\mathbf{X}}^{\Pi}$ *is the smallest topology on* $\mathbf{X} \subset \mathbf{Y}^{\mathbf{F}}$ *such that* π_f *is continuous for each* $f \in \mathbf{F}$.

$\mathcal{T}_{\mathbf{X}}^{\Pi}$ coincides with the topology $\mathcal{T}_{\mathbf{X}}^{\mathbf{F}}$ generated (see Definition 2.4.34) on the space $\mathbf{X}$ by the family $\{\pi_f \mid f \in \mathbf{F}\} \longleftrightarrow \mathbf{F}$ of the functions $f : x \in \mathbf{X} \subset \mathbf{Y}^{\mathbf{F}} \mapsto f(x)$.

$\mathcal{T}_{\mathbf{X}}^{\Pi} \equiv \mathcal{T}_{\mathbf{X}}^{\mathbf{F}}$ has as a base the family of sets

$$\left\{x \in \mathbf{X} \,\middle|\, f_1(x) \in O_1, \ldots, f_k(x) \in O_k\right\}$$

for arbitrary $k \in \mathbb{N}$, $f_1, \ldots, f_k \in \mathbf{F}$, and $O_1, \ldots, O_k \in \mathcal{T}_{\mathbf{Y}}$.

Again, these definitions are independent of a pre-existing topology on $\mathbf{X}$, if any.

Definition 2.4.26. *A net* $\{x_\alpha\}_{\alpha \in D}$, $x_\alpha \in \mathbf{X}$ ***converges*** *to* $x \in \mathbf{X}$ *with respect to* $\mathcal{T}_{\mathbf{X}}^{\mathbf{F}}$ *whenever the nets* $\{f(x_\alpha)\}_{\alpha \in D}$ *converge to* $f(x)$ *for each* $f \in \mathbf{F}$. *The topology* $\mathcal{T}_{\mathbf{X}}^{\mathbf{F}}$ *is also called* **F**- ***topology***, *or* ***weak topology*** *for* $\mathbf{X} \subset \mathbf{Y}^{\mathbf{F}}$.

Definition 2.4.27. ***w-convergence*** *is the convergence relative to a weak topology* $\mathcal{T}_{\mathbf{X}}^{\mathbf{F}}$. *In symbols:*

$$x_\alpha \rightharpoonup x \quad \text{whenever} \quad f(x_\alpha) \to f(x) \quad \text{for each} \quad f \in \mathbf{F}.$$

Let $\mathbf{F}$ be a family of functions from a set $\mathbf{X}$ to a uniform space $(\mathbf{Y}, \mathcal{V})$. Then the topology $\mathcal{T}_{\mathbf{F}}^{\mathbf{X}}$ of pointwise convergence is the topology of a uniformity:

Definition 2.4.28. ***Uniformity of pointwise convergence*** $\mathcal{U}_{\mathbf{F}}$ *is the product uniformity on the space* $\mathbf{Y}^{\mathbf{X}}$ *relativized to its subset* $\mathbf{F}$, *i.e., such that each of the relativized projections* $\pi_x : \mathbf{F} \to \mathbf{Y}$ *is uniformly continuous, relative to* $\mathcal{U}_{\mathbf{F}}$.

The following results hold true.

a) $\mathcal{U}_{\mathbf{F}}$ has as a subbase the sets $W_x(V)$ defined, for each $V \in \mathcal{V}$ and $x \in \mathbf{X}$, by

$$W_x(V) := \Big\{(f,g) \in \mathbf{F} \times \mathbf{F} \Big| (f(x), g(x)) \in V\Big\}; \qquad (2.4.8)$$

b) $\mathcal{T}_{\mathcal{U}_{\mathbf{F}}}$ coincides with the topology $\mathcal{T}_{\mathbf{F}}^{\mathbf{X}}$ of pointwise convergence;

c) $\{f_\alpha\}_{\alpha \in D}$ is a Cauchy net in $(\mathbf{F}, \mathcal{U}_{\mathbf{F}})$ if and only if $\{f_\alpha(x)\}_{\alpha \in D}$ is a Cauchy net for each $x \in \mathbf{X}$.

Let now $W(V)$, for $V \in \mathcal{V}$, denote the set of the form

$$W(V) := \Big\{(f,g) \in \mathbf{F} \times \mathbf{F} \Big| (f(x), g(x)) \in V, \quad \forall x \in \mathbf{X}\Big\}. \qquad (2.4.9)$$

Definition 2.4.29. ***Uniformity*** $\mathcal{U}_U$ ***of uniform convergence*** *is the uniformity for* $\mathbf{F} \subset \mathbf{Y}^{\mathbf{X}}$ *which has as a base the family of the sets* $W(V)$ *for* $V \in \mathcal{V}$.

Definition 2.4.30. ***Topology of uniform convergence*** *is the topology* $\mathcal{T}_{\mathcal{U}_U}$ *of the uniformity* $\mathcal{U}_U$.

Clearly $W(V) \subset W_x(V)$. This implies that $\mathcal{U}_{\mathbf{F}}$ (and $\mathcal{T}_{\mathcal{U}_{\mathbf{F}}}$) are smaller, i.e., coarser, than $\mathcal{U}_U$ (and $\mathcal{T}_{\mathcal{U}_U}$).

The following results hold true.

a) $\mathcal{U}_U$ is generated by the family of all pseudo-metrics of the form

$$d_\infty(f,g) := \sup_{x \in \mathbf{X}} \{d(f(x), g(x))\}$$

where d is a bounded member of the gage $G_{\mathcal{V}}$. Boundedness is without loss of generality.

b) A net $\{f_\alpha\}_{\alpha\in D}$, $f_\alpha \in \mathbf{F}$, ***converges uniformly*** to $g \in \mathbf{F}$ when it converges relative to the topology $\mathcal{T}_{\mathcal{U}_U}$. This happens when the net is Cauchy relative to $\mathcal{U}_U$ and pointwise converging (i.e.: for each $x \in \mathbf{X}$ the net $\{f_\alpha(x)\}_{\alpha\in D}$ converges to $g(x)$). Equivalently: whenever for each $V \in \mathcal{V}$ there exists $\alpha_0 \in D$ such that $\alpha \geq \alpha_0$ implies $f_\alpha(x) \in V[g(x)]$ for all $x \in \mathbf{X}$.

c) If $(\mathbf{Y}, \mathcal{V})$ is complete, so it is the uniform space $(\mathbf{F}, \mathcal{U}_U)$.

Definition 2.4.31. ***Topology*** $\mathcal{T}_{\mathcal{U}_R}$ ***of uniform convergence on the sets*** $R \in \mathcal{R}$, *where* $\mathcal{R}$ *is any family of subsets of* $\mathbf{X}$, *is the topology of the uniform convergence on the members of* $\mathcal{R}$. *The topology* $\mathcal{T}_{\mathcal{U}_R}$ *has as subbase the family of sets of the form*

$$\left\{ f \in \mathbf{F} \;\middle|\; \sup_{x\in R} d(f(x), g(x)) < \delta, \quad d \in G_{\mathcal{V}},\; g \in \mathbf{F},\; R \in \mathcal{R},\; \delta > 0 \right\}.$$

Convergence relative to $\mathcal{T}_{\mathcal{U}_R}$ means uniform convergence on each $R \in \mathcal{R}$. In particular, if $\mathcal{R} \equiv \{\mathbf{X}\}$ one has the topology of uniform convergence; whereas, if $\mathcal{R}$ is the collection of all singletons $\{x\}$ one has the topology of pointwise convergence. In particular:

Definition 2.4.32. ***Topology*** $\mathcal{T}_{\mathcal{U}_K}$ ***of uniform convergence on the compact sets*** $K \subset \mathbf{X}$ *is the topology on the set* $\mathbf{F} \subset \mathbf{Y}^{\mathbf{X}}$ *of the functions* $f : \mathbf{X} \to (\mathbf{Y}, \mathcal{V})$ *of the uniformity* $\mathcal{U}_K$ *that has as a subbase the family of sets of the form*

$$W_K(V) := \left\{ (f, g) \;\middle|\; (f(x), g(x)) \in V \quad \text{for all } \; x \in K \text{ compact} \right\}.$$

It is clear that $\quad \mathcal{T}_{\mathbf{F}}^{\mathbf{X}} \equiv \mathcal{T}_{\mathcal{U}_{\mathbf{F}}} \subset \mathcal{T}_{\mathcal{U}_K} \subset \mathcal{T}_{\mathcal{U}_U}$.

Definition 2.4.33. *A family* C *of functions* $f : \mathbf{X} \to (\mathbf{Y}, \mathcal{V})$ *is said to be* ***equicontinuous*** *when, for each* $x \in \mathbf{X}$ *and for each* $V \in \mathcal{V}$, *there exists a neighborhood* $\mathcal{N}(x)$ *such that*

$$f(\mathcal{N}(x)) \subset V[f(x)] \qquad \text{for all } \; f \in C.$$

Equivalently:

$$\bigcap_{f \in \mathbf{F}} f^{-1}\left(V[f(x)]\right) \quad \text{is a neighborhood of } x \text{ for each } V \in \mathcal{V}.$$

Equivalently: when for each $x \in \mathbf{X}$, $d_{\mathcal{V}} \in G_{\mathcal{V}}$, and $s > 0$, there exists $\mathcal{N}(x)$ such that: if $x' \in \mathcal{N}(x)$ and $f \in C$ then $d_{\mathcal{V}}(f(x'), f(x)) < s$.

Equivalently: whenever the convergence to $x \in \mathbf{X}$ of a net $\{x_\alpha\}_{\alpha \in D}$, $x_\alpha \in \mathbf{X}$, implies the convergence, ***uniform with respect to*** $f \in C$, of each net $\{f(x_\alpha)\}_{\alpha \in D}$ to the point $f(x) \in \mathbf{Y}$.

The $\mathcal{T}_C^{\mathbf{X}}$-closure of an equicontinuous family C is also equicontinuous, and its members are continuous functions. If C is an equicontinuous family, then the topology $\mathcal{T}_C^{\mathbf{X}}$ of pointwise convergence coincides with the topology $\mathcal{T}_{\mathcal{U}_K}$.

The following theorem is fundamental. Let $\mathcal{C}(\mathbf{X}, \mathbf{Y})$ be the family of continuous function on a regular, locally compact, Hausdorff space $\mathbf{X}$ to a Hausdorff uniform space $(\mathbf{Y}, \mathcal{V})$, and let $\mathcal{C}$ be endowed with the topology $\mathcal{T}_{\mathcal{U}_K}$.

THEOREM 2.4.1. ***(Ascoli-Arzelà)*** *([KEa] Section 7.17)*
A set $C \subseteq \mathcal{C}(\mathbf{X}, \mathbf{Y})$ is relatively compact (i.e., has $\mathcal{T}_{\mathcal{U}_K}$ - closure which is compact) if and only if: C is equicontinuous and for each $x \in \mathbf{X}$ the set $C[x] := \{f(x) \mid f \in C\}$ has compact closure in $\mathbf{Y}$.

2.4.3 The linear spaces case

Firstly, let $\mathbf{F}$ denote a family of functionals $f : \mathbf{X} \to \mathbb{R}$ from a set $\mathbf{X}$ to $\mathbb{R}$.

Definition 2.4.34. ***Topology*** $\mathcal{T}_{\mathbf{X}}^{\mathbf{F}}$ ***generated by*** $\mathbf{F}$ *is the smallest topology on* $\mathbf{X}$ *such that each* $f \in \mathbf{F}$ *is continuous.*

The topology $\mathcal{T}_{\mathbf{X}}^{\mathbf{F}}$ may have as a subbase the following family of sets $\{f^{-1}(O_{\mathbb{R}}),\ O_{\mathbb{R}} \in \mathcal{T}_{|\cdot|}\}$, $f \in \mathbf{F}$, $\mathcal{T}_{|\cdot|} \subset 2^{\mathbb{R}}$.

Example 2.4.3. The topology $\mathcal{T}_C^{\Pi}$ of pointwise convergence for $C([a,b],\,\mathbf{R})$ coincides with the topology $\mathcal{T}_C^{[a,b]}$ generated on C by the set of functions $x \in [a,b]$ such that $x \equiv \pi_x : f \in C \mapsto f(x) \in \mathbf{R}$. Indeed, when each $\pi_x : C \to \mathbf{R}$ is continuous then $f_n \to f$ whenever $f_n(x) \to f(x)$ for $x \in [a,b]$, (in general, not uniformly).

□

The space $\left(\mathbf{X}, \mathcal{T}_{\mathbf{X}}^{\mathbf{F}}\right)$ is Hausdorff iff $\mathbf{F}$ separates the points of $\mathbf{X}$.

If the space $\mathbf{X}$ is already equipped with a topology $\mathcal{T}$ and is T1 and completely regular, then the set $C(\mathbf{X},\mathbf{R})$ of the continuous functionals is sufficiently rich ([ROa] Section 8.3) to have $\mathcal{T}_{\mathbf{X}}^{C}$ identical to $\mathcal{T}$.

Let now, in particular, $(\mathbf{V},\mathcal{T}), (\mathbf{V}_1,\mathcal{T}_1), (\mathbf{V}_2,\mathcal{T}_2)$ be linear topological spaces; let $\mathcal{L}(\mathbf{V}_1,\mathbf{V}_2)$ denote the space of all linear maps $f : \mathbf{V}_1 \to \mathbf{V}_2$; and, in particular, let $\mathbf{V}^* := \mathcal{L}(\mathbf{V},\mathbf{R})$ be the linear space of continuous linear functionals x^* on $\mathbf{V}$.

Definition 2.4.35. *A family* $\mathbf{F}$ *of functionals* $f : \mathbf{V} \to \mathbf{R}$ ***separates the points*** *of* $\mathbf{V}$ *when* $f(x) = 0$ *for all* $f \in \mathbf{F}$ *implies* $x = 0$.

Definition 2.4.36. ***Seminorm*** d *is an absolutely homogeneous, subadditive, functional on* $\mathbf{V}$.

Definition 2.4.37. ***Weak topology*** *on* $\mathbf{V}$ *is the topology* $\mathcal{T}_{\mathbf{V}}^{\mathbf{V}^*}$ *generated by the family* $\mathbf{V}^*$ *of all continuous linear functionals.*

Equivalently ([ASa] Theorem 3.5.2) $\mathcal{T}_{\mathbf{V}}^{\mathbf{V}^*}$ is the topology $\mathcal{T}_{\mathbf{V}}^{G^*}$ generated by (the gage G^* of) the seminorms d_{x^*} defined by $d_{x^*}(x) := |x^*(x)|$, for all $x^* \in \mathbf{V}^*$.

The topology $\mathcal{T}_{\mathbf{V}}^{\mathbf{V}^*}$ may be generated by the base of neighborhoods of the origin

$$\left\{ y \in \mathbf{V} \,\middle|\, |x_i^*(y)| < r, \quad i = 1,\ldots,k \right\}, \qquad \text{for} \quad k \in \mathbf{N}, \quad r > 0,$$

or, by the subbase of the sets

$$V_{d_{x^*},r}[x] := \left\{ y \in \mathbf{V} \;\middle|\; |x^*(x-y)| < r \right\}, \quad r > 0, \quad x^* \in \mathbf{V}^* .$$

It may be useful to recall that if $\mathbf{V}_1$ is normed and $\mathbf{V}_2$ is Banach then also $\mathcal{L}(\mathbf{V}_1, \mathbf{V}_2)$ is Banach. Moreover ([DUa] Section V 3.15) a linear mapping connecting two Banach spaces $T : \mathbf{V}_1 \to \mathbf{V}_2$ is continuous if and only if it is weakly continuous.

Definition 2.4.38. ***Isomorphism*** $\varsigma : \mathbf{V}_1 \to \mathbf{V}_2$ *is a* 1−1 *map in* $\mathcal{L}(\mathbf{V}_1, \mathbf{V}_2)$ *such that* $\varsigma \mathbf{V}_1 = \mathbf{V}_2$.

If $\mathbf{V}_1$ and $\mathbf{V}_2$ are normed then the isomorphism ς is also ***isometric*** if $|\varsigma x| = |x|$ for all $x \in \mathbf{V}_1$.

Definition 2.4.39. *A net* $\{x_\alpha\}_{\alpha \in D}$ *is said to be* ***weakly convergent****, or* ***convergent in the weak topology*** $\mathcal{T}_{\mathbf{V}}^{\mathbf{V}^*}$, *in a normed linear space* $\mathbf{V}$, *when there exists an* $x \in \mathbf{V}$ *such that the net* $x^*(x_\alpha)$ *converges to* $x^*(x)$ *for each* $x^* \in \mathbf{V}^*$.

The following is often used to guarantee w-convergence in linear spaces.

THEOREM 2.4.2. *A sequence* $\{x_k\}_{k \in \mathbf{N}}$, $x_k \in \mathbf{V}$ *normed linear, is* w*-convergent to* $x \in \mathbf{V}$ *if (and only if)*

a) $\sup_{k \in \mathbf{N}} |x_k| < \infty$;

b) $f(x_k) \to f(x)$ *for all* $f \in \mathbf{V}_0^*$, *where* $\mathbf{V}_0^*$ *is such that the smallest closed subspace of* $\mathbf{V}^*$ *that contains all the elements of* $\mathbf{V}_0^*$ *is* $\mathbf{V}^*$ *itself.*

For the space $C^*(\mathbf{X}, \mathbf{R})$ of the continuous functionals on the space $C(\mathbf{X}, \mathbf{R})$, where $\mathbf{X}$ is a Hausdorff compact space, measure theory and integration theory are used to prove

THEOREM 2.4.3. ***(Riesz representation)*** *([DUa] Chapter 6) If* $\mathbf{X}$ *is a Hausdorff compact space then* $C^*(\mathbf{X}, \mathbf{R})$ *is isometrically isomorphic with the space of regular bounded measures* μ *defined*

on the σ-algebra $\mathcal{B}$ of the Borel sets of $\mathbf{X}$. *The correspondence between f^* and μ verifies the relation*

$$f^*(f) = \int_B f \, d\mu \,, \qquad f \in \mathcal{C}, \quad B \in \mathcal{B}. \tag{2.4.10}$$

It follows ([KOa] Section 20.3) from this and from Lebesgue Dominated Convergence theorem that: uniform boundedness and pointwise convergence of a sequence $\{f_k\}_{k\in\mathbf{N}}$ of functions in $\mathcal{C}(\mathbf{X},\mathbf{R})$ are not only necessary, but also sufficient conditions for the sequence $\{f_k\}_{k\in\mathbf{N}}$ to be weakly convergent to $f \in \mathcal{C}(\mathbf{X},\mathbf{R})$.

Example 2.4.4. The $\widehat{\delta}$-functionals on $\mathcal{C}([a,b],\mathbf{R})$ coincide with the projections $\pi_x : f \to f(x)$, $f \in \mathcal{C}$, and belong to $\mathcal{C}^*$. Each $\widehat{\delta}$ may be seen as the weak limit, in $\mathcal{C}^*$, of a δ-sequence $\{\widehat{\delta}_{,k} \in \mathcal{C}^*\}_{k\in\mathbf{N}}$ (see Eq. (2.3.15)) such that

$$\widehat{\delta}_{,k} \rightharpoonup \widehat{\delta} \quad \text{when} \quad f(\widehat{\delta}_{,k}) \to f(\widehat{\delta}) \quad \forall f \in \mathcal{C}^{**} \,,$$

i.e., owing to the reflexibility of $\mathcal{C}([a,b],\mathbf{R})$, when $\widehat{\delta}_{,k}(f) \to \widehat{\delta}(f)$ $\forall f \in \mathcal{C}$. If a sequence of functions $\{\delta_{,k} \in \mathcal{C}\}_{k\in\mathbf{N}}$ is such that

$$\widehat{\delta}_{,k}(f) \equiv \int_a^b f(t)\;\delta_{,k}(t)\,dt$$

then it is possible to denote by δ the ***function*** satisfying

$$\int_a^b f(t)\;\delta_{,k}(t)\,dt \to \int_a^b f(t)\;\delta(t)\,dt := f(0) \qquad \text{as } k \to \infty \,.$$

Hence, to each $\widehat{\delta}$-functional one may correspond a δ-function which is the limit of a δ-sequence of functions $\{\delta_{,k} \in \mathcal{C}\}_{k\in\mathbf{N}}$ in the sense of the weak-limit of the corresponding functionals.

□

For $L_p(\mathbf{X}, \mathcal{S}, \mu)$ spaces the following may be proved ([DUa] Section IV 8.1).

THEOREM 2.4.4. *If $1 < p < \infty$ and $p^{-1} + q^{-1} = 1$ then there is an isometric isomorphism between $L_p^*(\mathbf{X}, \mathcal{S}, \mu)$ and $L_q(\mathbf{X}, \mathcal{S}, \mu)$. The correspondence between $f^* \in L_p^*$ and $g \in L_q$ verifies the relation*

$$f^*(f) = \int_S g\, f\, d\mu\,, \qquad f \in L_p(\mathbf{X}, \mathcal{S}, \mu)\,, \quad S \in \mathcal{S}\,. \tag{2.4.11}$$

In the case of $p = 1$, an additional assumption is needed ([DUa] Section IV 8.5).

THEOREM 2.4.5. *Let $(\mathbf{X}, \mathcal{S}, \mu)$ be a σ-finite measure space. Then there exists an isometric isomorphism between $L_1^*(\mathbf{X}, \mathcal{S}, \mu)$ and $L_\infty(\mathbf{X}, \mathcal{S}, \mu)$. Moreover, the correspondence between the points $f^* \in L_1^*$ and the points $b \in L_\infty(\mathbf{X}, \mathcal{S}, \mu)$ verifies the relation*

$$f^*(f) = \int_S b\, f\, d\mu\,, \qquad f \in L_1(\mathbf{X}, \mathcal{S}, \mu)\,, \quad S \in \mathcal{S}\,. \tag{2.4.12}$$

In this case, as we have seen above, for each $b \in L_\infty$ the function

$$d_b(f) := |f^*(f)| \equiv \left|\int_S b\, f\, d\mu\right|\,, \qquad f \in L_1(\mathbf{X}, \mathcal{S}, \mu)\,, \quad S \in \mathcal{S}\,, \tag{2.4.13}$$

is a semi-norm on $L_1(\mathbf{X}, \mathcal{S}, \mu)$.

Consequently, a sequence $\{f_n \in L_1(\mathbf{X}, \mathcal{S}, \mu)\}_{k \in \mathbf{N}}$ weakly converges to $f \in L_1(\mathbf{X}, \mathcal{S}, \mu)$ when for each $b \in L_\infty(\mathbf{X}, \mathcal{S}, \mu)$ one has

$$\int f_n\, b\, d\mu \longrightarrow \int f\, b\, d\mu \qquad \text{in } \mathbf{R}, \quad \text{as} \quad n \to \infty\,.$$

The space $(L_1(\mathbf{X}, \mathcal{S}, \mu), \mathcal{U}_{G_b})$ equipped by the uniformity $\mathcal{U}_{G_b}$ generated by the gage $G_b := \{d_b\}_{b \in L_\infty}$ is Hausdorff.

Let L_1^b denote the corresponding gage space, i.e., let d_b be defined as in (2.4.13) and

$$L_1^b := \left(L_1(\mathbf{X}, \mathcal{S}, \mu), \mathcal{T}_{\mathcal{U}_{G_b}}\right), \qquad G_b := \left\{d_b\right\}_{b \in L_\infty(\mathbf{X}, \mathcal{S}, \mu)}.$$

Finally, let $\mathbf{Y}$ be a regular locally compact space, and $\mathcal{C}(\mathbf{Y}, L_1^b)$ denote the topological space of the continuous functions $f : \mathbf{Y} \to L_1^b$ equipped with the topology $\mathcal{T}_{\mathcal{U}_K}$ of the uniform convergence on the compact subsets of $\mathbf{Y}$.

Then Ascoli-Arzelà theorem may be used on the subsets C of the (complete) space $\mathcal{C}\left(\mathbf{Y}, L_1^b\right)$.

2.5 References

[ANa] Antosik P., Mikusinski J., and Sikorski R., **Theory of Distributions**, Elsevier, Amsterdam, (1973).

[ASa] Ash R. B., **Real Analysis and Probability**, Academic Press, New York, (1970).

[DUa] Dunford N. and Schwartz J. T. **Linear Operators**, J. Wiley, London, (1988).

[HAa] Haloms P.R., **Measure Theory**, Van Nostrand, Princeton N.J., (1965).

[KEa] Kelley L., **General Topology**, GTM 27, Springer, Heidelberg, (1975).

[KGa] Kogan M., **Rarefied Gas Dynamics**, Plenum Press, New York, (1969).

[KOa] Kolomgorov A. N., and Fomin S. V., **Introductory Real Analysis**, Dover, New York, (1975).

[LOa] Loeve M., **Probability Theory**, GTM 46, Springer, Heidelberg, (1978).

[ROa] ROYDEN L., **Real Analysis**, MacMillan, New York, (1968).

[RUa] RUELLE D., **Statistical Mechanics, Rigorous Results**, W.A. Benjamin, Reading, (1969).

Chapter 3

Models of Population Dynamics with Stochastic Interactions

3.1 Introduction

The class of models dealt with in this chapter concerns the population dynamics of individuals subject to interactions of kinetic-type. Models of this class are characterized by a mathematical structure similar to those of the phenomenologic kinetic theory, and in particular by the Boltzmann equation. Indeed, the evolution equations are derived following a line quite close to that used for the Boltzmann equation, and they also consist of nonlinear systems of integro-differential equations for the probability densities over the ***states*** of each population.

The first model of this type was proposed by Jager and Segel [JAa] to study the evolution of the physical state relative to certain populations of insects. Experimental analysis on the actual behavior of such populations is due to Hogeweg and coworkers [HOa]. Some developments and generalizations were studied in [ARa] with particular attention to the dynamics of several pairwise interacting populations and to the qualitative analysis of the initial value problem. Models with multiple interactions are proposed in [ARb], where the existence of equilibrium solutions is also studied. Models with space structure have been introduced in [BLa]. A discussion on the general framework on this subject may be found in the review paper [BLc].

As we have seen in Chapter 1, Jager and Segel model concerns the evolution of the probability density of a population of anonymous interacting organisms. To be more precise, if $u \in [0,1]$ represents a suitable variable called the ***dominance***, $t \in [0,T]$ the time and if

$$f : (t,u) \in [0,T] \times [0,1] \mapsto f(t,u) \in \mathbb{R}_+ \tag{3.1.1}$$

is a function such that $f(t,\cdot)$ belongs to $L_1([0,1];\mathbb{R}_+)$ for each $t \in [0,T]$, then $f = f(t,\cdot)$ is assumed to represent the probability density, with respect to the dominance u, of the population at the time t. Hence, f satisfies the normalization condition

$$\int_0^1 f(t,u)\,du = 1\,, \qquad \forall t \geq 0\,. \tag{3.1.2}$$

As we shall see, a nonlinear evolution equation for f is found by means of a suitable balance equation. In the case of binary encounters, as for the Boltzmann equation, the evolution equation is characterized by quadratic nonlinearities; in the case of triple encounters, nonlinearities are of cubic order. All models described in this chapter are somewhat related to the original Jager and Segel model. The general framework is that one of the mathematical theory of population dynamics [HPa].

The chapter is organized into five sections.

Section 3.1 is this introduction.

Section 3.2 contains the basic modelling of population dynamics referred to a system of several interacting populations of individuals subject to binary and multiple interactions.

Sections 3.3 and 3.4 are of analytic contents: Section 3.3 analyses the existence of solutions of the corresponding initial value problem; Section 3.4 the existence of equilibrium solutions.

Section 3.5 deals with some applications and provides a discussion on some possible generalizations of this class of models.

The organization of this chapter will somehow be repeated by the following chapters. Namely, the first part describes the class of models. Then, a survey follows the analytic results and applications available in the literature. Finally, a critical analysis of the models is proposed.

The analytic and computational problems related to the class of models dealt with in this chapter are far less complicated than those related to the Boltzmann model. The simplification is due not only to homogeneity of the spatial description, but mainly to the fact that the distribution f preserves L_1-norm. This property can naturally be exploited both towards existence proofs and solution techniques.

3.2 The Generalized Jager and Segel Model

As already mentioned, Jager and Segel [JAa] proposed a model suitable to describe the statistical evolution of the dominance factor in a population of several interacting organisms. ***Dominance*** is the ability of part of a population of certain insects, e.g., bombo bees, to organize the activity of the remaining part of the population. This behavior has experimentally been observed by Hogeweg [HOa].

Technically this model can be generalized to several interacting populations subject to multiple interactions. The model can even be developed to describe social interactions different from those which are the subject of the original paper. In this case, dealing with several populations and taking account of multiple collisions can be useful.

The axiomatization that follows is already related to generalized models. The assumptions which define the mathematical model are:

Assumption 3.2.1. *The physical system consists of n interacting populations. The physical state of each population is parametrized by a variable $u \in [0, 1]$, here in after called* ***state****.*

Assumption 3.2.2. *The (one-particle) probability distribution of* *each population admits a probability density function*

$$f_i \,:\, (t,u) \in [0,T] \times [0,1] \mapsto f_i(t,u) \in I\!R_+ \,, \tag{3.2.1}$$

for $i \in \{1,\dots,n\}$. *Hence the probability of finding, at time* t, *an individual of the* i*-th population in a state within the interval* $[u_1, u_2] \subseteq [0,1]$ *is given by*

$$P_i(t, u \in [u_1,u_2]) = \int_{u_1}^{u_2} f_i(t,u)\, du\,. \tag{3.2.2}$$

Assumption 3.2.3. *The number of pairwise encounters per unit time between individuals of the* i*-th population in the state* v *and individuals of the* j*-th population in the state* w *is assigned by the* ***encounter rate*** $\eta_{ij}^{(2)}(v,w) \geq 0$. *Similarly, the encounter rate of triple interactions among individuals of the* i*-th population in the state* v, *of the* j*-th population in the state* w *and of the* m*-th population in the state* z, *is given by* $\eta_{ijm}^{(3)}(v,w,z) \geq 0$.

Assumption 3.2.4. *The probability that an individual of the* i*-th population in the state* v *transits to the state* u *because of an encounter with an individual of the* j*-th population in the state* w *admits a* ***transition probability density****, with respect to the variable* u, *denoted by* $\psi_{ij}^{(2)}(v,w;u) \geq 0$. *The probability that after a triple encounter the element starting from the state* v *ends up in the state* u *has a density denoted by* $\psi_{ijm}^{(3)}(v,w,z;u) \geq 0$.

Assumption 3.2.5. *Only binary and triple encounters occur. The time derivative of each* f_i *does not depend on* u.

The evolution equation for density functions f_i is then obtained, as in the Boltzmann case, by equating the time derivative of f_i to the balance between ***gain*** and ***loss*** terms. In particular one can consider both the contribution of binary and triple interaction terms.

As mentioned, factorization of the joint probability is assumed, and modelling may follow the paths outlined in Section 1.2. Actually, factorization cannot be completely justified on a mathematical basis, and it has to be regarded as an approximation of physical reality. In fact, it can be accepted for limited time intervals and when the system refers to a large number of non distinguishable individuals, however it is relatively more critical in the case of multiple interactions.

As in the Boltzmann case, the model is based on assumptions that are stated on the microscopic interactions among individuals, and provides information about statistical, and hence macroscopic, variables.

The model's mathematical structure, in the case of binary and triple interactions, resides in the following set of n balance equations

$$\frac{\partial f_i}{\partial t} = G_i^{(2)}[\mathbf{f}] + G_i^{(3)}[\mathbf{f}] - L_i^{(2)}[\mathbf{f}] - L_i^{(3)}[\mathbf{f}], \qquad (3.2.3)$$

where $i \in \{1, \ldots, n\}$ and $\mathbf{f} = \{f_i\}_{i=1}^n$. The gain and loss terms have the following expressions

$$G_i^{(2)}[\mathbf{f}](t,u) = \sum_{j=1}^n \int_0^1 \int_0^1 \eta_{ij}^{(2)}(v,w)\psi_{ij}^{(2)}(v,w;u) f_i(t,v) f_j(t,w)\, dv\, dw,$$

$$G_i^{(3)}[\mathbf{f}](t,u) = \sum_{j=1}^n \sum_{m=1}^n \int_0^1 \int_0^1 \int_0^1 \eta_{ijm}^{(3)}(v,w,z)\psi_{ijm}^{(3)}(v,w,z;u)$$

$$\times f_i(t,v) f_j(t,w) f_m(t,z)\, dv\, dw\, dz,$$

$$L_i^{(2)}[\mathbf{f}](t,u) = f_i(t,u) \sum_{j=1}^n \int_0^1 \eta_{ij}^{(2)}(u,w) f_j(t,w)\, dw,$$

$$L_i^{(3)}[\mathbf{f}](t,u) = f_i(t,u) \sum_{j=1}^n \sum_{m=1}^n \int_0^1 \int_0^1 \eta_{ijm}^{(3)}(u,w,z)$$

$$\times f_j(t,w) f_m(t,z)\, dw\, dz\,. \tag{3.2.4}$$

The explicit mathematical model follows, for $i = 1, \ldots, n$,

$$\begin{aligned}
\frac{\partial f_i}{\partial t}(t,u) = \sum_{j=1}^{n} \bigg\{ \int_0^1 \int_0^1 \eta_{ij}^{(2)}(v,w)\psi_{ij}^{(2)}(v,w;u) f_i(t,v) f_j(t,w)\, dv\, dw \\
- f_i(t,u) \int_0^1 \eta_{ij}^{(2)}(u,w) f_j(t,w)\, dw \bigg\} \\
+ \sum_{j=1}^{n} \sum_{m=1}^{n} \bigg\{ \int_0^1 \int_0^1 \int_0^1 \eta_{ijm}^{(3)}(v,w,z) \\
\times \psi_{ijm}^{(3)}(v,w,z;u) f_i(t,v) f_j(t,w) f_m(t,z)\, dv\, dw\, dz \\
- f_i(t,u) \int_0^1 \int_0^1 \eta_{ijm}^{(3)}(u,w,z) f_j(t,w) f_m(t,z)\, dw\, dz \bigg\}.
\end{aligned} \tag{3.2.5}$$

With obvious meaning of symbols we have, in compact form,

$$\frac{\partial \mathbf{f}}{\partial t} = \mathbf{J}^{(2)}[\mathbf{f}] + \mathbf{J}^{(3)}[\mathbf{f}]\,, \tag{3.2.6}$$

where $\mathbf{J}^{(2)} = \{J_i^{(2)}\}_{i=1}^n$, $\mathbf{J}^{(3)} = \{J_i^{(3)}\}_{i=1}^n$, and

$$\begin{cases} J_i^{(2)}[\mathbf{f}](t,u) = G_i^{(2)}[\mathbf{f}](t,u) - L_i^{(2)}[\mathbf{f}](t,u)\,, \\ J_i^{(3)}[\mathbf{f}](t,u) = G_i^{(3)}[\mathbf{f}](t,u) - L_i^{(3)}[\mathbf{f}](t,u)\,. \end{cases} \tag{3.2.7}$$

The model represented by Eqs. (3.2.5), or in the compact form (3.2.6), has to be regarded as a generalization of Jager and Segel model, and it can effectively be applied to the analysis of some specific physical system only if the interaction terms η and ψ are specialized and properly identified for that particular system. On the other hand, although the identification of these two terms is a rather difficult task, some of the applications developed so far gave interesting and encouraging results. This will be seen in Section 3.5 where the analytic and computational treatment of the model will be referred to the general form (3.2.5), and the applications referred to specific models.

3.3 On the Initial Value Problem

This section deals with the qualitative analysis of the solutions of the mathematical problems opened by the models described above. Existence and uniqueness of the initial value problem solutions is proved with reference to the developments of computational methods. Existence of equilibrium points can also be proved by taking into account that all simulations showed a trend towards equilibrium. Only a detailed analysis of the stability of the said equilibrium points is at present not available in the literature, for general cases.

Consider the initial value problem for the mathematical model with binary and triple interactions

$$\begin{cases} \dfrac{\partial \mathbf{f}}{\partial t} = \mathbf{J}^{(2)}[\mathbf{f}] + \mathbf{J}^{(3)}[\mathbf{f}] \\ \mathbf{f}(t=0,u) = \mathbf{f}_0(u)\,. \end{cases} \tag{3.3.1}$$

The operators $\mathbf{J}^{(2)}$ and $\mathbf{J}^{(3)}$, defined in Section 3.2, will be assumed to be such that, for all $i, j, m = 1, \ldots, n$ and almost all $u, v, w, z \in [0,1]^4$, the following properties hold true:

- the terms η are real valued, measurable, and uniformly bounded functions, i.e., satisfying

$$0 \le \eta_{ij}^{(2)}(v,w) \le K_\eta^{(2)}\,, \qquad \text{and} \qquad 0 \le \eta_{ijm}^{(3)}(v,w,z) \le K_\eta^{(3)}\,, \tag{3.3.2}$$

  where $K_\eta^{(2)}$ and $K_\eta^{(3)}$ are real constants;

- the terms ψ are non-negative probability densities

$$0 \leq \psi_{ij}^{(2)}(v, w; u)\,, \qquad \text{and} \qquad \int_0^1 \psi_{ij}^{(2)}(v, w; u)\, du = 1\,, \quad (3.3.3)$$

$$0 \leq \psi_{ijm}^{(3)}(v, w, z; u)\,, \qquad \text{and} \qquad \int_0^1 \psi_{ijm}^{(3)}(v, w, z; u)\, du = 1\,. \quad (3.3.4)$$

We are interested in the existence of solutions. This analysis was first developed, see [ARa], for the Cauchy problem arising from models with binary interactions. Then, a generalization of the results to models with multiple interactions was proved in [ARb]. A brief survey of the analysis of [ARa], [ARb] is given in the following.

Suitable definitions of solutions, and of the Banach spaces wherein they are contained, need to be given in that existence and uniqueness of the initial value problem solutions are proved by using classical fixed point theorems and semigroup theory [BEa], [BEb].

Definition 3.3.1. *Denote by* $\mathbf{X}$ *the space of* n*-tuples* $\mathbf{g}$ *of real functions* $g : [0,1] \to \mathbf{R}$ *such that* $g \in L_1([0,1];\mathbf{R})$. *The space* $\mathbf{X}$ *is Banach if it is endowed with the norm*

$$\|\mathbf{g}\| = \sum_{k=1}^{n} \|g_k\| = \sum_{k=1}^{n} \int_0^1 |g_k|\, du\,. \quad (3.3.5)$$

Definition 3.3.2. *Denote by* $\mathbf{J} = \{J_i\}_{i=1}^n$, $\mathbf{G} = \{G_i\}_{i=1}^n$, *and* $\mathbf{L} = \{L_i\}_{i=1}^n$, *the* n*-tuples of operators from* $\mathbf{X}$ *into* $\mathbf{X}$ *are respectively defined by*

$$J_i[\mathbf{f}](t,\cdot) = J_i^{(2)}[\mathbf{f}](t,\cdot) + J_i^{(3)}[\mathbf{f}](t,\cdot)\,, \quad (3.3.6a)$$

$$G_i[\mathbf{f}](t,\cdot) = G_i^{(2)}[\mathbf{f}](t,\cdot) + G_i^{(3)}[\mathbf{f}](t,\cdot)\,, \quad (3.3.6b)$$

$$L_i[\mathbf{f}](t,\cdot) = L_i^{(2)}[\mathbf{f}](t,\cdot) + L_i^{(3)}[\mathbf{f}](t,\cdot)\,, \quad (3.3.6c)$$

for $i = 1, \ldots, n$, *and* $t \in [0, T]$. *In fact, the two operators* $\mathbf{G}$ *and* $\mathbf{L}$ *map into itself the closed positive cone* $\mathbf{X}^+$ *of* $\mathbf{X}$, *i.e., the subset of* $\mathbf{X}$ *of all* n*-tuples of non-negative valued functions* $g : [0, 1] \to \mathbb{R}_+$. *From Eq. (3.2.7) it follows that* $\mathbf{J}$ *can be written as the difference* $\mathbf{J} = \mathbf{G} - \mathbf{L}$.

Definition 3.3.3. *Denote by* $\mathcal{B} = \{B_i\}_{i=1}^n$ *the* n*-tuple of mappings from* $\mathbf{X}$ *into* $\mathbf{X}$ *that, for each* $t \in [0, T]$, *are defined by*

$$B_i[\mathbf{f}](t, u) = \sum_{j=1}^{n} \left\{ \int_0^1 \eta_{ij}^{(2)}(u, w) f_j(t, w)\, dw \right.$$
$$\left. + \sum_{m=1}^{n} \int_0^1 \int_0^1 \eta_{ijm}^{(3)}(u, w, z) f_j(t, w) f_m(t, z)\, dw\, dz \right\}, \quad (3.3.7)$$

for $i = 1, \ldots, n$, *and* $u \in [0, 1]$.

Remark 3.3.1. *Comparing Definitions 3.3.2 and 3.3.3 one may see that, for each* $\mathbf{f}(t, \cdot) \in \mathbf{X}^+$ *and* $t \in [0, T]$,

$$L_i[\mathbf{f}](t, \cdot) = f_i(t, \cdot)\; B_i[\mathbf{f}](t, \cdot), \qquad i = 1, \ldots, n. \quad (3.3.8)$$

Definition 3.3.4. *An* n*-tuple* $\mathbf{f} = \{f_i\}_{i=1}^n$ *of functions from* $[0, T]$ *to* $\mathbf{X}^+$ *is defined to be a* ***solution*** *of the initial value problem (3.3.1) when* $\mathbf{f}$ *maps* $[0, T]$ *into* $\mathbf{X}^+$, *it is continuously differentiable, and it satisfies on* $[0, T]$ *the system of* n *nonlinear differential equations*

$$\frac{\partial \mathbf{f}}{\partial t} = \mathbf{J}[\mathbf{f}] \equiv \mathbf{G}[\mathbf{f}] - \mathbf{f}\mathcal{B}[\mathbf{f}], \quad (3.3.9)$$

with initial conditions $\mathbf{f}_0(\cdot) := \mathbf{f}(t = 0, \cdot) \in \mathbf{X}^+$.

Existence of solutions for arbitrarily large time intervals is stated by the following

THEOREM 3.3.1. *If conditions (3.3.2)–(3.3.4) are satisfied, there exists a unique mapping* $\mathbf{f} : [0,T] \to \mathbf{X}^+$ *that is a solution of the initial value problem (3.3.1) and satisfies*

$$\|\mathbf{f}(t,\cdot)\| = \|\mathbf{f}_0\|, \quad \forall t \in [0,T]. \tag{3.3.10}$$

Proof: Let c be an arbitrary positive constant. It may be shown that there exist positive numbers

$$M_1 = M_1(c, n, K_\eta^{(2)}, K_\eta^{(3)}), \qquad M_2 = M_2(c, n, K_\eta^{(2)}, K_\eta^{(3)}) \tag{3.3.11}$$

such that, whenever $\mathbf{g}_1, \mathbf{g}_2 \in \mathbf{X}^+$ satisfy $\|\mathbf{g}_1\| \le c$ and $\|\mathbf{g}_2\| \le c$, one has

$$\|\mathbf{J}[\mathbf{g}_1]\| \le M_1, \quad \|\mathbf{J}[\mathbf{g}_2]\| \le M_1, \quad \|\mathbf{J}[\mathbf{g}_1] - \mathbf{J}[\mathbf{g}_2]\| \le M_2\|\mathbf{g}_1 - \mathbf{g}_2\|. \tag{3.3.12}$$

Thanks to these inequalities, if $\|\mathbf{f}_0\| \le \frac{1}{2}c$, one can find a t_0 depending only on c such that:

i) There exists a unique solution $\mathbf{f}$ to Problem (3.3.1) on $[0, t_0)$.

ii) The solution $\mathbf{f}$ verifies on $[0, t_0)$ the following set of equations

$$f_i(t,u) = f_{i0}(u) \exp\left(-\int_0^t B_i[\mathbf{f}](s,u)\, ds\right)$$
$$+ \int_0^t G_i[\mathbf{f}](s,u) \exp\left(-\int_s^t B_i[\mathbf{f}](r,u)\, dr\right) ds, \tag{3.3.13}$$

for $i = 1, \ldots, n$ and $u \in [0,1]$.

Equation (3.3.13) proves that the solution $\mathbf{f}$ is positive for positive initial conditions. Finally, since the properties of the terms ψ imply that

$$\|\mathbf{f}(t,\cdot)\| = \|\mathbf{f}_0\|, \qquad \text{for each} \qquad t \in [0, t_0), \tag{3.3.14}$$

the solution can be extended to an arbitrary time interval $[0, T]$ such that $0 < T < \infty$. ∎

Regularity of solutions can be obtained under suitable smoothness assumptions on the initial conditions. Existence of regular solutions can then be proved in a suitable subspace of $\mathbf{X}$.

Still referring to the spatially homogeneous case, a further generalization of the model consists in modifying the right-hand side of the evolution model by introducing suitable activation terms. In this case the model can be written as

$$\frac{\partial f_i}{\partial t} + \frac{\partial}{\partial u}\left(k_i(t, u) f_i\right) = J_i[\mathbf{f}]\,, \qquad \text{for} \quad i = 1, \ldots, n\,, \tag{3.3.15}$$

where the term $k_i = k_i(t, u)$ simulates the direct action over the distribution f_i.

The analysis of the initial value problem for Eq. (3.3.15) can be developed by analogous methods. Qualitative analysis can be developed after a detailed reconstruction of the properties of operator $\mathbf{J}$. Preliminary analysis of this problem can be based on semigroup theory [BEa], [BEb], and the treatment may be found in [ARc] for some particular specialization of the term $\mathbf{b}$. It requires several technical calculations and hence it will not be reported here. In addressing the interested reader to [ARc], we simply note that the overall procedure is standard: one writes the semigroup corresponding to the evolution problem and then shows its contractivity properties.

3.4 Stationary Points

In the analysis and validation of a population model, an important role is played by the study of existence and stability of equilibrium points. Indeed it may be interesting to show whether each interacting population has a trend towards a certain equilibrium configuration. This matter is well documented in classical population dynamics, see

[HPa], however only a few preliminary results are known for the Jager and Segel model which are limited to the case $b_i = 0, \ \forall i = 1, \ldots, n$.

Bearing this in mind, and with reference to [ARb], consider the stationary problem

$$\mathbf{J}[\mathbf{f}] = 0\,, \tag{3.4.1}$$

for $\mathbf{f}(t, \cdot) \equiv \mathbf{f}(\cdot)$ in the positive cone $\mathbf{X}^+$ of $\mathbf{X}$, and assume the following

Assumption 3.4.1. *For all $i, j, m \in \{1, \ldots, n\}$, the terms $\eta_{ij}^{(2)}$ and $\eta_{ijm}^{(3)}$ are positive constants, denoted by $c_{ij}^{(2)}$ and $c_{ijm}^{(3)}$ respectively.*

Assumption 3.4.2. *For all $i, j, m \in \{1, \ldots, n\}$, the real valued functions $\psi_{ij}^{(2)}$ and $\psi_{ijm}^{(3)}$ are continuous on $[0,1]^3$ and $[0,1]^4$, respectively.*

Under the above hypotheses, the stationary problem $\mathbf{J}[\mathbf{f}] = 0$ can be written in the equivalent form

$$\mathbf{f} = \mathbf{Q}\,[\mathbf{f}]\,, \qquad \mathbf{Q} \,:\, \mathbf{X}^+ \to \mathbf{X}^+\,, \tag{3.4.2}$$

where the operator $\mathbf{Q} := \{Q_i\}_{i=1,\ldots,n}$ is defined by

$$\begin{cases} Q_i[\mathbf{f}](\cdot) = \dfrac{G_i[\mathbf{f}](\cdot)}{B_i[\mathbf{f}](\cdot)} & \text{if} \quad \|\mathbf{f}\| > 0 \\ Q_i[\mathbf{f}](\cdot) = 0 & \text{if} \quad \|\mathbf{f}\| = 0 \end{cases} \tag{3.4.3}$$

for $\quad i = 1, \ldots, n$.

The following theorem proves the existence of equilibrium points.

THEOREM 3.4.1. *Under Assumptions 3.4.1 and 3.4.2, the operator $\mathbf{Q}$ maps on itself the closed bounded and convex subset of $\mathbf{X}^+$ defined by*

$$\mathcal{D} = \left\{ \mathbf{f} \in \mathbf{X}^+ \,:\, \int_0^1 f_i(u)\, du = 1\,, \quad \forall\, i = 1, \ldots, n \right\}. \tag{3.4.4}$$

Moreover $\mathbf{Q} : \mathcal{D} \to \mathcal{D}$ *is completely continuous. Hence, Schrauder fixed point theorem may be used to prove the existence of at least one stationary solution.*

The analysis refers to the necessary properties for the operator $\mathbf{Q}$ to verify Schrauder theorem. They are: $\mathbf{Q}(\mathcal{D}) \subset \mathcal{D}$ and $\mathbf{Q} : \mathcal{D} \to \mathcal{D}$ completely continuous. These properties are indeed possessed by the operator defined by Eq. (3.4.3), as it is proved by technical calculations documented in [ARb]. On the other hand, uniqueness and stability of stationary solution was confirmed by the simulations developed by Preziosi [PRa]. Stability of equilibrium points was separately proved by Rondoni [ROa], with the method of Boltzmann maps [STa]. This powerful technique allowed one to prove uniqueness and stability of equilibrium points by means of suitable Liapunov functionals. However, the proof required suitable assumption on the terms η and ψ, see [ROa]. A stability proof under more general conditions, such as those developed in [PRa], is not yet available. The only stability result [CAa] is given under special assumption on the term ψ, such that the evolution equation admits a Liapunov functional similar to the H-functional of the Boltzmann equation.

It is immediate to notice that if the quantities η and ψ are uniformly equal to one, then exponential stability is immediately proved. Indeed, it is easy to show that the solution $\mathbf{f}$, such that $f_i = 1$ for all i, is analytic and asymptotically exponentially stable. As already remarked a general proof of the stability of equilibrium points would be very useful in the applications, see e.g., [HOa]. Indeed, among the various general aspects of the problem, it is of interest to show whether interacting populations may or may not reach, asymptotically in time, any equilibrium configurations.

3.5 Applications and Perspectives

The mathematical model described in this chapter defines a general framework for several other models that will be proposed in the chapters that follow. The model (in fact, the class of models) has been described in great generality without any reference to specific applications. This section shows how this model can be technically used to describe some interesting physical systems. Then, some perspectives will be discussed.

This section is divided into four subsections. The first one deals with a general methodology; it allows the modelling of the encounter rate η and transition probability density ψ defined in Assumptions 3.2.3 and 3.2.4. The second subsection describes the details of a mathematical model applied to social sciences. The third one proposes a model in mathematical epidemiology. The last one deals with a discussion on research perspectives, with special attention to possible developments in areas different from the afore-mentioned ones.

3.5.1 Modelling η and ψ

This subsection analyses some methodological aspects that concern the modelling of terms η and ψ. The idea commonly followed consists in modelling the above quantities by means of simple analytic expressions, characterized by a few parameters to be identified by comparison with experimental data. Clearly, this analysis is preliminary to the development of any specific model.

A general way to model the ***encounter rate*** η is the following

$$\eta_{ij}(v,w) = \eta^0_{ij}\ \phi(v,w)\,, \tag{3.5.1}$$

where η^0_{ij} is a constant depending only on the types of interacting populations, and $\phi = \phi(v,w)$ is a dimensionless function of the interacting pair states. For instance, concerning the term η^0 one may assume that encounters are allowed or inhibited on the basis of their belonging to a selected set of pairs among all the possible ones. Thus one has

$$\begin{cases} \eta_{ij}^0 = c_{ij}\,, & \text{if} \quad (i,j) \in I_*\,, \\ \eta_{ij}^0 = 0\,, & \text{if} \quad (i,j) \notin I_*\,, \end{cases} \tag{3.5.2}$$

where I_* is the prescribed subset of admissible pairs of indices, and where the transition coefficients matrix is symmetric: $c_{ij} = c_{ji}$. A typical example of this encounter law is represented by a model that regards the encounters within individuals of the same population to be more frequent than those between individuals of different populations.

Concerning the term ϕ, one may assume that ***short range interactions***, i.e., those between individuals that are in states close to each other, are more frequent than those between individuals in distant states. A technical example is the following

$$\begin{cases} \phi(v,w) = 1\,, & \text{if} \quad |v-w)| \le d \le 1\,, \\ \phi(v,w) = 0\,, & \text{if} \quad |v-w| > d\,. \end{cases} \tag{3.5.3}$$

Clearly, other possibilities are acceptable as well. For instance, one may assume a smooth reduction of the mobility, i.e., that the interaction frequency gradually decreases when the distance increases between the interacting pair states

$$\phi(v,w) = \frac{1}{1 + c(v-w)^2}\,, \qquad c > 0\,; \tag{3.5.4}$$

or that high states produce social isolation

$$\phi(v,w) = \frac{1}{1 + c_1 v + c_2 w}\,, \qquad c_1, c_2 > 0\,. \tag{3.5.5}$$

The above examples give simple ideas about some conceivable modelling. Additional examples can be developed following lines of the same kind.

Identification of the ***transition probability density*** ψ is then required. As mentioned, it consists of a two-parameter family of probability densities $\psi = \psi(v, w; u)$ with respect to the variable u. On the other hand, it is assumed that for each choice of the parameters v, w the correspondent density function $u \mapsto \psi(v, w; u)$ may be characterized by its first and second moments, i.e., by the expectation value $m := E[u] \equiv E_1[\psi(v, w; \cdot)]$ of the post interaction state u, and by its standard deviation $\sigma := E[(u - m)^2]$. Clearly, both these quantities, $m = m(v, w)$ and $\sigma = \sigma(v, w)$, are considered to be functions of the type and states of the interacting pair. Then, in particular, ψ is formally written as follows

$$\psi_{ij}(v, w; u) = \psi_{ij}\left(u; m_{ij}(v, w), \sigma_{ij}(v, w)\right) \tag{3.5.6}$$

and for all $i, j \in \{1, \ldots, n\}$, and $v, w \in [0, 1]$, one has

$$\int_0^1 \psi_{ij}\left(u; m_{ij}(v, w), \sigma_{ij}(v, w)\right) du = 1\,. \tag{3.5.7}$$

As is well known, several examples are available, in probability theory, the explicit form of a family of densities with the cited properties could be used to assess ψ when m and σ are experimentally identified. A relatively simple case occurs when both the above quantities have a constant value, independent of the interacting pair features. For instance, if the output is deterministically identified then ψ can be assessed as a delta distribution function of the type

$$\psi_{ij}\left(u; m_{ij}(v, w)\right) = \delta\left(u - m_{ij}(v, w)\right)\,. \tag{3.5.8}$$

On the other hand, when a sufficiently large amount of different experimental data is available, one may identify the densities using a larger number of parameters. For instance, orthogonal expansions of the probability density can be used, as shown in Chapter 4 of [BLb] where the expansion coefficients are related to the experimental values of the moments.

As mentioned, the above identification is based on the assumption that the quantity m is constant with time. Suitable models, such as those we will show in the second one of the following applications, assume that the most probable outcome also evolves with time.

3.5.2 A model of social behaviors

As a first application we consider a model of ***social behavior*** among individuals of a single population. In this model the variable u is linked to the ***social state***, and the aim is to simulate the behavior of individuals that belong to several social classes. Pairwise interactions are allowed either within the same class or among different classes. Interactions may modify the social state of the interacting individuals.

This first, very simple, application is developed with tutorial aims to show the basic aspects of the modelling. In details, the model is defined by the following assumptions:

Assumption 3.5.1a. *All individuals are indistinguishable and belong to one single population. Only binary encounters are significant.*

Assumption 3.5.2a. ***Egoistic behavior*** – ***Model I*** – Interac*tions are such that an encounter between individuals of different social states determines a further increase of the higher social state, and decrease of the lower one. The transition probability density is characterized by a constant value for the variance, and by a most probable output state given by*

$$\begin{cases} m(v,w) = v - \alpha v(1-w)\,, & \text{if} \quad v \le w\,, \\ m(v,w) = v + \alpha(1-v)w\,, & \text{if} \quad v \ge w\,, \end{cases} \tag{3.5.9}$$

where $\alpha \in [0,1]$. The encounter rate η may be either assumed to be independent of the states of the interacting individuals, or to decay when the state distance increases.

Assumption 3.5.3a. ***Altruistic behavior – Model II*** – *Interactions are such that an encounter between individuals of different social states determines an averaging of the social states. The transition probability density is characterized by a constant value for the variance, and by a most probable output state given by*

$$m(v, w) = v + \alpha(w - v), \qquad \alpha \in [0, 1]. \tag{3.5.10}$$

The encounter rate η is assumed to be independent of the states of the interacting individuals.

Assumption 3.5.4a. ***Socially dialectic behavior – Model III*** *– Interactions are such that an encounter between individuals of different social states determines an altruistic behavior if the distance between the states of the interacting individuals is larger than a certain critical value, while a competitive behavior arises if the distance is below the above critical value. In the first case the encounter rate η is assumed to decay when the state distance increases; in the second case it is assumed to be independent of the interacting distance.*

Simulation of the model consists in solving the initial value problem and computing the time evolution of the probability density f. Quantitative results can be obtained by collocation methods as described in Chapter 4 of [BLb]. The method is technically simple: it consists in interpolating the probability density $f = f(t, u)$ as follows

$$f(t, u) \cong f^m(t, u) = \sum_{h=1}^{m} \chi_h(u) f_h(t), \tag{3.5.11}$$

where $f_h(t) = f(t, u_h)$ corresponds to the collocation

$$\{u_1 = 0, \dots u_h, \dots, u_m = 1\}, \tag{3.5.12}$$

and where the terms χ_h, for $h = 1, \ldots, m$, denote a set of fundamental functions such that

$$\chi_h(u_k) = \delta_{hk} , \tag{3.5.13}$$

where δ_{hk} is the Kronecker symbol. As known, the above property is satisfied, among others, by Lagrange polynomials, by fundamental splines, by Sinc functions, and so on.

Inserting the interpolation–approximation (3.5.11) into the evolution equation, and replacing the integrals by weighted sums, yields a system of m ordinary differential equations on the components $f_h = f_h(t)$. Standard methods may then be used to solve these equations equipped with suitable initial conditions, and hence the evolution of the nodal functions f_h is obtained. Finally, interpolation (3.5.11) provides the approximate evolution of f.

In the case of the problem above, simulation has been organized for different initial conditions. Several experiments showed an asymptotic trend towards an equilibrium configuration. In particular, the following features have been observed.

a) In the case of Model I, simulation shows that the social dynamics concentrates on the states also goes towards their extreme values.

b) In the case of Model II, social dynamics leads to an idealistic concentration of states around the middle zone.

c) In the case of Model III, social dynamics leads to a distribution which is still focused mainly on the central zone, however a socially controlled positive concentration is allowed in the extreme zones.

d) The asymptotic distribution appears to be stable in all simulation, although depending on initial data.

The above trend towards equilibrium further motivates the analytic study of equilibrium points stability described in Section 3.4.

Further specialization may also be obtained. For instance, one may deal with several interacting populations. Then, social interactions may include social conflicts etc.

3.5.3 A model in epidemiology

The mathematical model described in Section 3.5.2 has been applied to epidemiology first in [PRa]. Here we describe a model aimed to describe a population of carriers of parasites, and that contains an activation term such as that included in Eq. (3.3.15).

Consider a population of non isolated individuals that are ***carriers*** of a pathologic state, e.g., of ***parasites***. Individuals interact among themselves and exchange parasites at each contact. Moreover, parasites proliferate on each individual until they are released into the environment as soon as their number reaches a certain ***critical value*** u_c which is related to the maximum amount of parasites that may be carried by each individual. The assumptions that formalize the model (which will be called ***Model IV***) are the following:

Assumption 3.5.1b. *All individuals are indistinguishable and belong to one single population. Each individual is a carrier of parasites. The amount of parasites, i.e., the* ***infection degree*** *of the carrier, is identified by a variable* $u \in [0, 1]$. *When* u *reaches the critical value* $u_c = 1$, *the exceeding parasites are instantaneously released into the environment.*

Assumption 3.5.2b. *Only binary encounters are significant. The encounter rate,* η, *is a constant both with respect to time and to the states of the interacting individuals.*

Assumption 3.5.3b. *An exchange of parasites occurs whenever an encounter takes place between two individuals. Infection spreads from the individual with the higher infection degree to the individual with the lower one. The most probable post interaction amount of parasites, on the individual starting from the state* v *and that has been subject to an encounter with an individual in the state* w, *is assumed to depend both on the difference between the infection degrees:* $(w - v)$, *and on the proliferation of parasites during the time,* $1/\eta$, *that is expected to pass until the next encounter takes place.*

The above dependence follows the scheme

$$m(v, w) = u(1/\eta; u_0)\,,$$

where η is the encounter rate, and where $u = u(1/\eta; u(0))$ is the solution, with initial datum $u_0 = u(t = 0)$, of a suitable proliferation equation that rules the parasites evolution and that has the form:

$$\dot{u} = g(t, u)\,. \tag{3.5.14}$$

The state $u(0)$ is attained by the v-individual immediately after the encounter, and is assumed to be given by

$$u(0) = v + \alpha(w - v)\,, \tag{3.5.15}$$

where $0 \leq \alpha \leq 1$ denotes the degree of parasites transmission.

Assumption 3.5.4b. *Parasites naturally proliferate and die. Natural growth is proportional to the degree of infection u through a parameter $\beta \geq 0$ that represents the per-capita reproduction rate. Natural death is ruled by a coefficient $\gamma \geq 0$. Carriers release parasites into the environment as soon as u reaches the critical value $u_c = 1$. The degree of infection of an individual between two successive encounters, i.e., the parasites population, varies according to:*

$$\frac{du}{dt} = g(t, u) := \beta\ u - \gamma\,. \tag{3.5.16}$$

According to the above proliferation rule, the parasites population at a time t between two successive interactions is given by

$$u(t, u_0) = u_0\ e^{\beta\, t} + \frac{\beta}{\eta}\ \left[1 -\ e^{\beta\, t}\right]\,. \tag{3.5.17}$$

and hence the most probable post interaction amount m of the infection degree is

$$m(v,w) = \max\Big\{0\,,\ \min\{1\,,\ m^*(v,w)\}\Big\}\,, \tag{3.5.18}$$

where

$$m^*(v,w) = \Big[v + \alpha(w-v)\Big]\exp\left(\frac{\beta}{\eta}\right) + \frac{\beta}{\eta}\left[1 - \exp\left(\frac{\beta}{\eta}\right)\right]\,. \tag{3.5.19}$$

The time interval $1/\eta$ may be seen as a mean free time between two encounters. The minimum between the value m^* and 1 has to be taken because the total infection is not supposed to become larger than 1.

The above models are characterized by the fact that ψ is a probability density over u for all inputs v and w. In a more realistic picture, one may expect that individuals die when they reach the critical state $u = 1$ which represents their physiological limit.

3.5.4 Perspectives

The models proposed in this chapter can be regarded as ***generalized kinetic models***. They define the time evolution of population density functions over the individuals state variable u.

However, the mathematical structure of the class of models dealt with in this chapter is remarkably general, and can be successfully adjusted to represent physical systems quite different from those we have seen here. The same structure is shared by several of the models that will be proposed in the following chapters.

The amount of information provided by this class of models is larger than that delivered by traditional models of population dynamics. Indeed here the evolution is given by the probability density for each of the values of the physical variables that define the individuals state. Hence, a microscopic description, as that provided by deterministic models of population dynamics, may not be given. In the traditional models of population dynamics, indeed, the (continuous) state variable spaces are *a priori* subdivided into finite sets of states. Here, the state variables maintain their continuous character.

To understand this aspect one can think of the first one of the two models described in this section, namely the model of social behavior. When, *a posteriori*, we split the continuous distribution of the social state into, say, three states: the lower one $u_1 = 0$, the middle one $u_2 = .5$, and the higher one $u_3 = 1$, the evolution equation becomes a simple dynamical system with three populations: f_1, f_2, f_3 corresponding to u_1, u_2, u_3 respectively, and the evolution is described by a system of ordinary differential equations. Interactions modify the social state in a fashion that $f_1 + f_2 + f_3 = 1$.

In principle, one may also include interactions that are able to generate birth or death processes as well as changes of states. In this way, this type of models will retain features both of traditional models of population dynamics and of models with kinetic interactions.

The link between models with continuous distribution over the state u, stated in terms of integrodifferential equations, and models with discrete values of u, stated in terms of ordinary differential equations, is discussed in the last chapter of this book. Here we conclude this section with an analysis of the development perspectives that are generated by the class of models dealt with in this chapter.

We concentrate, in particular, on the characterization of the activation term b and on the developments of models with variable number of individuals. The analysis is kept at an introductory level, also considering that some of the above generalizations will be discussed in the next chapters.

The term b needs to be related to the variable u. For instance, if u is a scalar, as in the example dealt with in Section 3.5.4, then b is an activation term which models an action over the individuals of the population. On the other hand, if $\mathbf{u}$ is a multidimensional variable which includes the space variable, then the left-hand side term may include a space structure.

Modelling dynamics in space may be done by considering microscopic interactions between individuals as is done in [BLa] again referring to populations of insects. The model constructs a map of the type

$$(\mathbf{v}', \mathbf{v}'_w) = T(\mathbf{v}, \mathbf{v}_w)\,, \tag{3.5.20}$$

that relates precollision velocities $(\mathbf{v}, \mathbf{v}_w)$ to postcollision ones. The model developed in [BLa] is strictly related to the dominance factor and hence to the model by Jager and Segel. Dominants do not change their velocities after the interaction, whereas dominates reverse their velocities after an encounter with a dominant.

An alternative to the above kinetic approach is the introduction of a diffusion term as is suggested in [ARc]. The diffusion term may be linear or nonlinear. The evolution problem substantially differs from the ones we have seen in this chapter since boundary conditions have to be implemented.

A feature of the class of models dealt with in this chapter is that interactions do not generate new individuals nor cause their destruction. An important development consists in including the above effect. In Chapter 5 this type of models will be discussed and the utility of this generalization will be documented.

3.6 References

[ARa] ARLOTTI L. and BELLOMO N., Population dynamics with stochastic interactions, *Transp. Theory Stat. Phys.*, **24** (1995), 431–443.

[ARb] ARLOTTI L. and BELLOMO N., Population dynamics with stochastic interactions, *Appl. Math. Letters*, **9** (1996), 65–70.

[ARc] ARLOTTI L., BELLOMO N., and LACHOWICZ M., From Jager and Segel model to kinetic population dynamics, *Mathl. Comp. Modelling*, **30** (1999), 15–40.

[BEa] BELLENI-MORANTE A., **Applied Semigroup and Evolution Equations**, Clarendon Press, New York, (1980).

[BEb] BELLENI-MORANTE A. and MCBRIDE A.C., **Applied Nonlinear Semigroups**, Wiley, New York, (1998).

[BLa] BELLOMO N. and LACHOWICZ M., Mathematical biology and kinetic theory: Evolution of the dominance in a population of interacting organisms, in **Nonlinear Kinetic Theory and Mathematical Aspects of Hyperbolic Systems**, Boffi V. et al. Eds., World Scientific, London, Singapore, (1992), 11–20.

[BLb] BELLOMO N. and PREZIOSI L., **Modelling, Mathematical Methods and Scientific Computation**, CRC Press, Boca Raton (FL), (1995).

[BLc] BELLOMO N. and LO SCHIAVO M., From the Boltzmann equation to generalized kinetic models in applied sciences, *Mathl. Comp. Modelling*, **26** (1997), 43–76.

[CAa] CARAFFINI G., IORI M. and SPIGA G., On the connection between kinetic theory and a statistical model for the distribution of dominance in populations of social organisms, *Riv. Mat. Univ. Parma*, **5** (1996), 160–181.

[HOa] HOGEWEG P. and HESPER P., The ontogeny of the interaction structure in bumble bee colonies, *Behav. Ecol. Sociobiol.*, **12** (1983), 271–183.

[HPa] HOPPENSTEAD F., Mathematical Theory of Populations: Demographics, Genetics and Epidemics, *SIAM Conf. Series*, **2**, (1975).

[JAa] JAGER E. and SEGEL L., On the distribution of dominance in a population of interacting anonymous organisms, *SIAM J. Appl. Math.*, **52** (1992), 1442–1468.

[LOa] LONGO E., TEPPATI G. and BELLOMO N., Discretization of continuous models by Sinc interpolation methods, *Comp. Math. Appl.*, **32** (1996), 65–81.

[MAa] MacDonald n., **Biological Delay Systems**, Cambridge Univ. Press, Cambridge, (1992).

[PRa] Preziosi L., From population dynamics to modelling the competition between tumors and immune system, **23** (1996), 135–152.

[ROa] Rondoni L., Boltzmann-like kinetic models and Boltzmann maps, *Mathl. Comp. Modelling*, **25** (1997), 53–67.

[STa] Streater R.F., **Statistical Dynamics**, Imperial College Press, London, (1995).

Chapter 4

Generalized Kinetic Models for Coagulation and Fragmentation

4.1 Introduction

Coagulation fragmentation phenomena concern systems of a large number of particles which are allowed, on a statistical basis, to meet and coalesce to form larger "spherical" aggregates of compound radia and correspondingly compound masses. The aggregates are metastable in that they may not only proceed in joining with other individuals, and hence form clusters of still larger sizes: ***coagulation processes***, but may also undergo a destructive behavior, and break-up into individuals of smaller size: ***fragmentation processes***. In most cases, the system is assumed to have an infinite reservoir of identical (small, unitary) masses, so that the coagulation procedure may apriori proceed freely.

The class of models described in this chapter can be referred to several fields of applied sciences, and simulate phenomena wherein collisions and interactions between different individuals may modify the size of the interacting subjects. This type of phenomena, i.e. coagulation, fragmentation, and vaporization, can occur in molecular clusters, droplets, moist bubbles, reacting polymers, or even in aggregates of molecules in biology or natural sciences. Some of these applications will be discussed in the following chapters.

Modelling such phenomena may be done in a discrete framework [BAa] as well as in a continuous one. Continuous coagulation fragmentation phenomena may be described, as proposed by Smoluchowski, by a model which can be directly related to the class of models depicted in the first introductory chapter although some technical differences are clearly recognizable.

Denote by

$$c \;:\; (t,u) \in [0,T] \times \mathbb{R}_+ \;\mapsto\; c(t,u) \in \mathbb{R}_+ \tag{4.1.1}$$

the dependent variable $c = c(t,u)$ characterizing this class of models; it defines the number density function (with respect to the variable u and referred to a unit volume) of clusters of size u at time t. The model is an evolution equation for the variable c, that can formally be written as follows

$$\frac{\partial c}{\partial t} = J[c]\,. \tag{4.1.2}$$

The zeroth order moment with respect to the distribution c

$$E_0[c](t) = \int_0^\infty c(t,u)\,du \tag{4.1.3}$$

represents the total number of clusters (per unit volume) at time t; the first order moment

$$E_1[c](t) = \int_0^\infty u\; c(t,u)\,du\,. \tag{4.1.4}$$

is related to the average size (per unit volume). The mean value of the clusters size is

$$E[u](t) = \frac{E_1[c](t)}{E_0[c](t)}\,. \tag{4.1.5}$$

Both the continuous and the discrete models are based on two separate, independent, processes. Of these, one corresponds to the coalescence phenomenon K, the other to the fragmenting F. Each of them is responsible for a (time dependent) source and loss terms for the population of individuals of each possible size.

It is notable, however, that problems may arise in connection with the total mass of the system. Natural assumption is that $M_1(t)$ should be a constant, but it is shown [KRa] that this may not be the case, and ***gelation phenomena*** can occur unless certain convenient limited-growth conditions are imposed on the coalescence process.

Collisions may also be similar to those of the Boltzmann equation, even if one cannot realistically expect elastic collisions. In principle, a Boltzmann-like kinetic equation may be used to model a ***mixture of clusters*** linked to further coagulations or fragmentations. Vaporization phenomena may also be included, although not directly related to collisions.

The class of models reported in this chapter refers to the simplest among these phenomena; some of the above mentioned generalizations are discussed in the last section of the chapter. We remark that this class of models is proposed in the spatially homogeneous case, and that a conceivable development, particularly useful for the applications, may be that one of modelling space diffusion phenomena.

The chapter is divided into four sections.

Section 4.1 is this introduction.

Section 4.2 provides a description of the mathematical model.

Section 4.3 deals with a review of analytic results on the qualitative theory of the initial value problem.

Section 4.4 provides a survey on the applications of the model to the analysis of specific systems, and a discussion on some perspective developments of the model.

4.2 Description of the Models

Consider a system of ***particle aggregates***, or ***clusters***, homogeneously moving in space. Each cluster is characterized by its ***size*** $u = u(t)$. The ***state*** of the system is defined by a function $c = c(t, u)$ such that $c(t, u)du$ is the number of clusters, per unit volume, that at time t have size in du centered at $u \in \mathbb{R}$. Actually, the term cluster will be used in a general framework, which also includes droplets, cellular aggregates etc.

The variable u here is assumed to be continuous although in several physical systems, e.g. molecular clusters, it may assume only discrete values. Interactions may modify the size u in that loss and gain in the population of clusters of that size may happen.

An evolution model can be proposed with the same structure reported in Eq. (4.1.2). In particular, we consider the following

$$\frac{\partial c}{\partial t} = J[c] := G[c] - L[c] \,, \tag{4.2.1}$$

where G and L respectively represent the gain term into the state u, and the loss term from the state u. They account for the number of clusters that, respectively, gain and lose the state u in the unit time and unit volume.

Modelling the system consists in proposing explicit analytic expressions for the above mentioned terms. In the literature they are based on the following auxiliary quantities which, as is seen, are based on a two-body interaction assumption.

$K(u, w)$ is the rate at which clusters of size u coalesce with clusters of size w to form clusters of size $u + w$.

$F(u, w)$ is the rate at which clusters of size $u + w$ break up into two clusters of sizes u and w.

Each of the above terms generates a gain term and a loss term, that will be denoted by G_K, L_K and G_F, L_F, respectively. Their formal expressions are the following

$$G_K[c](t,u) = \frac{1}{2}\int_0^u K(u-w,w)c(t,u-w)c(t,w)\,dw\,, \quad (4.2.2)$$

$$L_K[c](t,u) = c(t,u)\int_0^\infty K(u,w)c(t,w)\,dw\,, \quad (4.2.3)$$

$$G_F[c](t,u) = \int_0^\infty F(u,w)c(t,u+w)\,dw\,, \quad (4.2.4)$$

$$L_F[c](t,u) = \frac{1}{2}\,c(t,u)\int_0^u F(u-w,w)\,dw\,. \quad (4.2.5)$$

The model, hence characterized by linear and quadratic terms, can be written explicitly

$$\begin{aligned}\frac{\partial c}{\partial t}(t,u) &= G_K(t,u) - L_K(t,u) + G_F(t,u) - L_F(t,u)\\ &= \frac{1}{2}\int_0^u K(u-w,w)c(t,u-w)c(t,w)\,dw\\ &- c(t,u)\int_0^\infty K(u,w)c(t,w)\,dw\\ &+ \int_0^\infty F(u,w)c(t,u+w)\,dw\\ &- \frac{1}{2}\,c(t,u)\int_0^u F(u-w,w)\,dw\,. \end{aligned} \quad (4.2.6)$$

Considering that this class of models can be related to the generalized kinetic models described in Chapter 1, it is worth discussing some of the main differences between the two classes of models. Essentially, we need to analyze the physical meaning of functions K and F.

Notice that only the term K is generated by binary encounters; on the contrary, F refers to the evolution of a single aggregate. In details:

- Binary encounters refer only to coalescence phenomena; they are described by the term $G_K - L_K$;
- Single evolutions refer to fragmentation activity; they are described by the term $G_F - L_F$.

This crude simplification allows to compare directly, with the models described in Chapters 1 and 3, only those models that are characterized by pure coalescence phenomena. In this case the evolution equation becomes

$$\frac{\partial c}{\partial t}(t,u) = \frac{1}{2}\int_0^u K(u-w,w)c(t,u-w)c(t,w)\,dw$$

$$-c(t,u)\int_0^\infty K(u,w)\,c(t,w)\,dw\,, \tag{4.2.7}$$

here the loss term is analogous to that of the models of Chapter 3, while the gain term loses one of the integrations, due to the fact that aggregation occurs only for certain sizes.

To make the model even more similar to those classified as generalized kinetic and population dynamics models, we can express the function K as a product of an encounter rate and a transition probability density

$$K(u-w,w) =: \eta(u-w,w)\psi(u-w,w;u)\,, \tag{4.2.8}$$

where:

$\eta(u-w,w)$ is the encounter rate between clusters of size $(u-w)$ and w.

$\psi(u-w,w;u)$ is the probability density that a cluster of size $(u-w)$ becomes a cluster of size u after an encounter, with coalescence, with a cluster of size w.

The function F is a breaking rate, and does not depend on η. Thus it is an inner property of the cluster.

Remark 4.2.1. *The details of functions K and F, and hence of η, ψ and F, must be fixed according to the particular physical process that is being modelled. These terms are generally assumed to be non-negative symmetric functions on $[0,\infty) \times [0,\infty)$. Actually, the splitting reported in Eq. (4.2.8) can contribute to the identification of these terms.*

Hence, the pure coalescence model can be written as

$$\frac{\partial c}{\partial t}(t,u) = \frac{1}{2}\int_0^u \eta(u-w,w)\psi(u-w,w;u)c(t,u-w)c(t,w)\,dw$$

$$-c(t,u)\int_0^\infty \eta(u,w)\psi(u,w;u+w)\,c(t,w)\,dw\,. \qquad (4.2.9)$$

In principle, one may design a model based on the assumption, similar to what is used in population dynamics, that the phenomenon is described by certain assigned binary interactions. In this case the model would be even more similar to the Boltzmann-like population models described in Chapter 3, and written as follows

$$\frac{\partial c}{\partial t}(t,u) = \int_0^\infty \int_0^\infty \eta(v,w)\psi(v,w;u)c(t,v)c(t,w)\,dv\,dw$$

$$-c(t,u)\int_0^\infty \eta(u,w)\psi(v,w;u)c(t,w)\,dw\,, \qquad (4.2.10)$$

where η and ψ are defined as above and aposteriori related to the kernels F and K. However, the physical interpretation of the above terms is not yet explicitly available in the literature.

4.3 Mathematical Problems

The analysis of the simulations provided by the model is related to the qualitative and quantitative behavior of the solutions of the following initial value problem

$$\begin{cases} \dfrac{\partial c}{\partial t} = J[c] \,, \\ c(0, \cdot) = c_0(\cdot) \,, \end{cases} \tag{4.3.1}$$

where the explicit form of $J[c](t, u)$ is given by the right-hand side of equation (4.2.6).

As usual, the analysis has to be developed bearing in mind the possible applications of the model towards real physical system simulations. Therefore the mathematics, which needs suitable assumptions on the terms K and F, should be developed in a framework sufficiently broad to include interesting physical assumptions. In other words, the assumptions on the terms K and F, need to be realistic from the physical point of view.

The problems that have been studied in the literature, can be classified into the following main classes:

- Existence of solutions of the initial value problem, and analysis of their asymptotic behavior;
- Existence of equilibrium solutions, stability and asymptotic time trend to equilibrium;
- Development of mathematical methods towards the computational solution of the initial value problem.

Although several mathematicians have been involved in the above outlined problems, starting from the pioneer paper by White [WHa], we will essentially refer, for the qualitative analysis, to the contents of the papers by Stewart [STa]-[STe]. Actually, all relevant bibliography is therein cited. On the other hand, quantitative analysis is developed in [AZa].

4.3.1. Existence of solutions

Existence of solutions in suitable Banach spaces can be developed by methods that are similar to those used in treating the Cauchy problem for the Boltzmann equation. The analysis developed by Stewart [STa] is based on compactness arguments. Here, a survey of the definitions and main results is given.

Let $\mathbf{X}$ be the space given by

$$\mathbf{X} := \left\{ y \in L_1\left([0,\infty]\right) \,\Big|\, \|y\| < \infty \right\}, \tag{4.3.2}$$

where

$$\|y\| := \int_0^\infty (1+u)|y(u)|\,du\,;$$

denote by $\mathbf{X}^+$ the closed positive cone defined by

$$\left\{ y \in \mathbf{X} \,\Big|\, y(u) \geq 0 \quad \lambda\text{-a.e.} \right\}.$$

The measure λ is the Lebesgue measure on the Borel algebra $\mathcal{B}(\mathbb{R})$.

The following definition is provided in [STa]

Definition 4.3.1. *Let $0 \leq \tau \leq \infty$. A solution of (4.3.1) on $[0,\tau)$ is a mapping $c : [0,\tau) \to \mathbf{X}^+$ such that*

a) $c(t,u) \geq 0$ *for all* $t \in [0,\tau)$ *and* $u \in [0,\infty)$;

b) $c : t \mapsto c(t,\cdot)$ *belongs to* $C^0([0,\tau))$ *and has values in* $\mathbf{X}^+$;

c) *the following bounds hold true*

$$\int_0^t \int_0^\infty K(u,w)c(s,w)\,dw\,ds < \infty\,, \tag{4.3.3a}$$

$$\int_0^t \int_0^\infty F(u,w)c(s,u+w)\,dw\,ds < \infty\,; \tag{4.3.3b}$$

d) *for all $t \in [0, \tau)$ the following is satisfied*

$$c(t,u) = c_0(u) + \int_0^t J[c](s,u)\,ds\,. \tag{4.3.3c}$$

The definition is well posed in the sense that: for almost all $u \in [0,\infty)$ the function $t \mapsto c(t,\cdot)$ given by (4.3.3c) is absolutely continuous on $[0,\tau)$, and (consequently) it may satisfy (4.3.1) for almost all $t \in [0,\tau)$.

Existence of solutions is proved for kernels K and F that are functions $\mathbb{R}^2 \to \mathbb{R}_+$ continuous, symmetric, possibly unbounded, although subject to the following growth conditions:

$$\begin{cases} K(u,w) \le \phi(u) + \phi(w) \quad \text{for all} \quad u, w \in [0,\infty)\,, \\ 0 \le \phi(u) \le k_1(1+u)^\alpha \quad \text{for some } \alpha \in [0,1), \quad k_1 \in \mathbb{R}\,, \\ F(u,w) \le k_2(1+u+w)^\beta \quad \text{for some } \beta \in [0,1), \quad k_2 \in \mathbb{R}. \end{cases} \tag{4.3.4}$$

THEOREM 4.3.1. *Under the above assumptions for kernels K and F, equation (4.3.1) has a solution $c : [0,\infty) \to \mathbf{X}^+$ for any $c_0 \in \mathbf{X}^+$.*

The proof undergoes several steps; the main of which are here sketched.

Step 1. Cut-off kernels K^n and F^n for $n = 1, 2, \ldots$ are defined, equal to the given K and F inside, and zero outside, the triangles

$$T_n := \left\{ (u,w) \in \mathbb{R}^2 \,\middle|\, 0 \le u, \quad 0 \le w, \quad u + w \le n \right\}. \tag{4.3.5}$$

Then, using standard contraction mapping techniques, the existence of a unique ***mass conservative solution***, in the sense of Definition (4.3.1),

$$c^n \in C^0\left([0,\infty),\ L_1\left([0,n]\right)\right) \tag{4.3.6}$$

is proved for each reduced equation

$$\frac{\partial c^n}{\partial t} = J^n[c^n]\,, \qquad c^n(0,u) = c_0^n(u)\,, \tag{4.3.7}$$

where $c_0^n \in \mathbf{X}^+$ equals c_0 for $u \leq n$ and is zero for $u > n$; and where J^n is defined by the right-hand side of equation (4.2.6) wherein the kernels are the cut-off ones.

Step 2. Each $c^n(t,\cdot)$ is extended to be zero outside $[0,n]$. Then, the sequence $\{c^n(t,\cdot)\}_{n\in\mathbb{N}}$ is seen to be a weakly relatively compact subset of $L_1([0,\infty])$ for each $t \in [0,\tau]$, $\tau < \infty$. Indeed, assumptions (4.3.4) allow convenient apriori estimates on the averages

$$\int_A c^n(t,u)\,du\,, \qquad A \in \mathcal{B}(\mathbb{R})\,,$$

on account of which Dunford-Pettis theorem [EDa] guarantees the claimed weak relative compactness of $\{c^n(t,\cdot)\}_{n\in\mathbb{N}}$ in $L_1([0,\infty])$.

This last is then used in the sense of relative compactness of $\{c^n(t,\cdot)\}_{n\in\mathbb{N}}$ in the space L_1^b, where (see Chapter 2)

> L_1^b is the space $L_1([0,\infty])$ equipped with the topology $\mathcal{T}_b$ induced by the gage of pseudometrics given by
>
> $$d_b(x,y) := \left| \int_0^\infty b(u)\,[x(u) - y(u)]\;du \right|\,, \tag{4.3.8}$$

where $b \in L_\infty([0,\infty])$.

Step 3. The topological space $C\left([0,\tau], L_1^b\right)$ is introduced, endowed with the topology $\mathcal{T}_{\mathcal{U}_{[0,\tau]}}$ of uniform convergence on the compact $[0,\tau]$, $\tau < \infty$.

Then, for each $t_0 \in [0,\tau]$, $\epsilon > 0$, and $b \in L_\infty([0,\infty])$, a neighborhood $\mathcal{N}(t_0)$ is seen to exist such that, if $t \in \mathcal{N}(t_0)$, then for all $n \in \mathbb{N}$ the following holds

$$d_b\left(c^n(t,\cdot), c^n(t_0,\cdot)\right) := \left| \int_0^\infty b(u)\left[c^n(t,u) - c^n(t_0,u)\right] du \right| < \epsilon . \tag{4.3.9}$$

This implies that: $\{c^n\}_{n\in\mathbb{N}}$ is an equicontinuous family in the topological space $\mathcal{C}\left([0,\tau], L_1^b\right)$.

Step 4. Steps 2 and 3 allow the use of Ascoli-Arzelà theorem (see again Chapter 2), which guarantees that the family $\{c^n\}_{n\in\mathbb{N}}$ is a relatively compact subset of $\mathcal{C}\left([0,\tau], L_1^b\right)$. Hence at least a subsequence

$$\{c^{n_h}\}_{h\in\mathbb{N}} \subseteq \{c^n\}_{n\in\mathbb{N}}$$

exists converging to a point $c \in \mathcal{C}\left([0,\tau], L_1^b\right)$.

As already mentioned, this is equivalent to saying that for $h \to \infty$ one has

$$c^{n_h}(t,\cdot) \rightharpoonup c(t,\cdot), \qquad \text{in } L_1\left([0,\infty]\right), \quad \text{uniformly on} \quad [0,\tau], \tag{4.3.10}$$

where we used the notation of weak convergence: $x^n \rightharpoonup x$ in L_1 whenever $x^*(x^n) \to x^*(x)$ in $\mathbb{R}$ for each continuous linear functional $x^* : L_1 \to \mathbb{R}$.

It has to be remarked, however, that this solution $c : t \mapsto c(t,\cdot)$ is not necessarily mass-conservative, although it is shown that $c(t,\cdot)$ belongs to $\mathbf{X}^+$, for each $t \in [0,\tau]$, as well as all the $c^n(t,\cdot)$ do.

Step 5. Let $t \in [0,\tau]$, $\tau < \infty$; and assume that $y(t,\cdot)$, and $y^n(t,\cdot)$ for $n = 1, 2, \ldots,$ are functions in $\mathbf{X}^+$. It is proved that if kernels K and F satisfy assumptions (4.3.4), and if

$$y^n(t,\cdot) \rightharpoonup y(t,\cdot), \qquad \text{in } L_1\left([0,u_0]\right), \qquad \text{as } n \to \infty . \tag{4.3.11}$$

Then for each $u_0 > 0$ and $t \in [0,\tau]$ the following holds

$$J^n[y^n](t,\cdot) \rightharpoonup J[y](t,\cdot), \qquad \text{in } L_1\left([0,u_0]\right), \qquad \text{as } n \to \infty . \tag{4.3.12}$$

In particular, functions $c(t,\cdot)$ and $c^n(t,\cdot)$ of equation (4.3.10) satisfy, for each $t \in [0,\tau]$, the weak convergence condition

$$J^n[c^n](t,\cdot) \rightharpoonup J[c](t,\cdot), \qquad \text{in } L_1([0,u_0]), \qquad \text{as } n \to \infty. \tag{4.3.13}$$

Step 6. Boundedness of the time integrals, guaranteed by assumptions (4.3.4), imply that as $n \to \infty$

$$\int_0^t J^n[c^n](s,\cdot)\,ds \rightharpoonup \int_0^t J[c](s,\cdot)\,ds\,, \qquad \text{in } L_1([0,u_0])\,, \tag{4.3.14}$$

and since equation (4.3.7) yields

$$c^n(t,\cdot) = c_0^n(\cdot) + \int_0^t J^n[c^n](s,\cdot)\,ds\,.$$

Uniqueness of the right-hand sides of (4.3.10) and (4.3.14), and arbitrariness of τ, finally prove the theorem.

□

Uniqueness of solutions is proved, in [STb], with standard techniques and under assumptions similar to (4.3.4). They are

$$\begin{cases} K(u,w) \le \phi(u)\cdot\phi(w) & \text{for all } u,w \in [0,\infty)\,, \\ 0 \le \phi(u) \le k_3(1+u)^{1/2} & \text{for some } k_3 \in \mathbb{R}\,; \\ \displaystyle\int_0^u F(u-w,w)(1+w)^{1/2}\,dw \le k_4(1+u)^{1/2} & \\ \qquad\qquad \text{for some } k_4 \in \mathbb{R}, \quad \text{and all } u \ge 0. & \end{cases} \tag{4.3.15}$$

The question of mass conservation is then arised in [STc,d] to extend a preceding discussion [ERa] that showed that Eq. (4.3.1) admits solutions that, at least in the pure coalescence case, are not necessarily mass conservative (***gelation***).

For pure coagulation processes the following result is proved.

THEOREM 4.3.2. *Assume that* $K(u,w) \geq \phi(u) \cdot \phi(w)$ *for a function* ϕ *such that* $\phi(u)/u \to \infty$ *as* $u \to \infty$. *Then no solution* $t \mapsto c(t,\cdot)$ *satisfying initial data* $c_0 \in \mathbf{X}^+$ *may exist on* $[0,\tau]$ *and be mass conservative if the function*

$$\exp(\delta\gamma_m) \int_m^\infty u \, c_0(u) \, du$$

does not tend to zero as $m \to \infty$, *for arbitrary* $\delta > 0$, *and where*

$$\gamma_m := \inf_{u \geq m} \left(\frac{\phi(u)}{u} \right) .$$

Further, a non mass-conservative solution is explicitly produced, under less restrictive assumptions, showing that mass conservation is not implied by existence. As well, for pure fragmentation processes, a non mass-conservative solution is shown to exist provided that

$$F(u,w) = (1+u+w)^{-\alpha} \qquad \text{for all } u,w \geq 0, \quad \text{and some } \alpha \in [0,1) .$$

Conversely, [STd], in the case of pure coagulation processes, a sufficient condition for the solutions to be mass conservative is that the kernel $K(u,w)$ be upper bounded by a linear function $k_5(u+w)$, $k_5 \in \mathbb{R}$, whenever u and w are greater than a certain $u_0 > 0$.

4.3.2. Equilibrium solutions and stability

Analysis of equilibrium solutions stability is clearly a relevant matter in interpretating the model towards applications. This subject is developed by Stewart [STe] using conservation equations and Lyapunov functional methods. In particular, convergence towards equilibrium solutions of equation (4.3.1) is studied in the restricted case of constant kernels, that is, for $K, F \in \mathbb{R}$.

With the same notation used in subsection 4.3.1, the equation now reads

$$\begin{aligned}\frac{\partial c}{\partial t}(t,u) =& \frac{1}{2}\int_0^u K\ c(t,u-w)c(t,w)\,dw \\ &- \int_0^\infty K\ c(t,u)c(t,w)\,dw + \int_0^\infty F\ c(t,u+w)\,dw \\ &- \frac{1}{2}\int_0^u F\ c(t,u)\,dw\,. \end{aligned} \tag{4.3.16}$$

The main result resides in the following

THEOREM 4.3.3. *Let $K, F \in \mathbb{R}$ be the constant kernels in equation (4.3.16), with $\lambda := F/K$. Let $c_0(u) \in \mathbf{X}^+$ be such that*

$$M_2 := E_2[c_0] = \int_0^\infty u^2 c_0(u)\,du\ <\ \infty\,, \tag{4.3.17}$$

and call

$$M_1 := E_1[c_0] = \int_0^\infty u\ c_0(u)\,du\,. \tag{4.3.18}$$

Then the equilibrium solution $u \mapsto \bar{c}(u)$ of Eq. (4.3.16) satisfying initial data $c(0,u) = c_0(u)$ is unique in $L_1([0,\infty])$ and given by

$$u \mapsto \bar{c}(u)\ :=\ \lambda \exp\left(-u\sqrt{\lambda/M_1}\right)\,. \tag{4.3.19}$$

Moreover, if a solution c of equation (4.3.16) satisfies initial data c_0 and the following conditions:

a) $\ln c(t,u)$ *exists for almost all* $u \geq 0$ *and is such that*

$$c(t,\cdot)(\ln c(t,\cdot) - 1) \quad \text{belongs to} \quad L_1([0,\infty]) \quad \text{for all} \quad t \geq 0\,;$$

b) $\varphi \in L_1([0,\infty])$ *and* $\delta > 0$ *exist such that for each* $t \geq 0$ *and* $|h| < \delta$

$$|(\ln c(t+h,u)-1)c(t+h,u)-(\ln c(t,u)-1)c(t,u)| \leq |h|\varphi(u)\,, \tag{4.3.20}$$

then, weakly in $L_1([0,\infty])$,

$$c(t,\cdot) \rightharpoonup \bar{c}(\cdot), \qquad \text{as} \quad t \to \infty\,,$$

where $\bar{c}$, *defined in Eq. (4.3.19), is the unique equilibrium solution of Eq. (4.3.16).*

As before the proof goes through various steps, after one has conveniently remarked that the hypotheses of Theorem 4.3.1 are in particular satisfied by those of Theorem 4.3.3.

Step 1. Under assumption (4.3.17), the second moment Eq. (4.3.16)

$$E_2[c](t) := \int_0^\infty u^2\, c(t,u)\, du$$

is proved to be bounded on $[0,\infty)$. This fact is used to deduce that

$$E_1[c](t) := \int_0^\infty u\, c(t,u)\, du \;=\; \int_0^\infty u\, c_0(u)\, du \;=\; M_1$$

and that

$$E_0[c](t) \;:= \int_0^\infty c(t,u)\, du \;\longrightarrow\; \sqrt{\lambda M_1} \qquad \text{as} \quad t \to \infty\,.$$

Hence this last needs to be the total number of clusters (per unit volume) of any equilibrium solution.

Step 2. A telescopic estimate of the mean distance between two stationary solutions of (4.3.16) is stated, which implies that any equilibrium solution of (4.3.16) is unique, in $L_1([0,\infty])$, for any $c_0 \in \mathbf{X}^+$. Then the equilibrium solution is easily recognized to be given by (4.3.19).

Step 3. The positive semiorbit

$$O^+(c_0) := \bigcup_{t\geq 0} \{c(t,\cdot)\}\,, \qquad c(0,\cdot) = c_0 \in \mathbf{X}^+\,,$$

of any solution of Eq. (4.3.16) that satisfies initial data: $c_0 \in \mathbf{X}^+$ and $E_2[c](0) =: M_2 < \infty$, is proved to be a weakly relatively compact set of $L_1([0,\infty])$. As in Step 2 of Theorem 4.3.1, this is done with the help of Dunford-Pettis theorem.

Then, the (weak) ***omega-limit set of*** c_0 is introduced, defined by

$$\omega(c_0) := \left\{ f \in L_1([0,\infty]) \,\middle|\, \exists\ \{t_n\}_{n\in\mathbb{N}}\,, \quad t_n \xrightarrow[\infty]{n} \infty, \quad c(t_n,\cdot) \rightharpoonup f(\cdot) \right\} \tag{4.3.21}$$

and continuity of solutions of (4.3.16) with respect to time is used to prove that

(3.a) $\omega(c_0)$ is non-empty;

(3.b) $c(t,\cdot) \rightharpoonup \omega(c_0)$, in $L_1([0,\infty])$, as $t \to \infty$;

(3.c) $\omega(c_0)$ is positively invariant, i.e. for each $f_0 \in \omega(c_0)$ the solution $f : t \mapsto f(t,\cdot)$ satisfying initial data f_0 has values $f(t,\cdot)$ that are in $\omega(c_0)$ for all $t \geq 0$. (In fact, owing to the results in Step 1, the set $\omega((c_0))$ is considered as a subset of $\mathbf{X}^+$.)

Step 4. The functional $V : y \in \mathbf{X}^+ \longmapsto \mathbb{R}_+$ is defined by

$$V[y] := \int_0^\infty \left(\ln \frac{y(u)}{\lambda} - 1 \right) y(u)\, du\,. \tag{4.3.22}$$

Then, the proof is sketched as follows

THEOREM 4.3.4. *Functional V is a weakly lower semicontinuous Lyapunov functional on $\mathbf{X}^+$ for the solutions $t \mapsto c(t,\cdot)$ of equation (4.3.16) which satisfy arbitrary initial data $c_0 \in \mathbf{X}^+$ and assumptions (a) and (b) above. That is*

a) $V[c(t,\cdot)](t) \le V[c_0(\cdot)](0) \in \mathbb{R}_+$ *for all* $t \ge 0$;

b) *if* $\{c^n(t,\cdot)\}_{n\in\mathbb{N}}$ *is in* $\mathbf{X}^+$ *for all* $t \ge 0$, *and if* $c^n(t,\cdot) \rightharpoonup c(t,\cdot)$ *in* $L_1([0,\infty])$ *as* $n \to \infty$ *then*

$$V[c(t,\cdot)](t) \le \liminf_{n\to\infty} V[c^n(t,\cdot)](t)\,. \tag{4.3.23}$$

Step 5. Let $c : t \mapsto c(t,\cdot)$ be a solution of Eq. (4.3.16) satisfying initial data $c_0 \in \mathbf{X}^+$ and all the assumptions of the theorem. Then, weak lower semicontinuity of functional V, non emptiness of $\omega(c_0)$, and again continuity with respect to time of solutions of Eq. (4.3.16) are used to prove that V is invariant over $\omega(c_0)$, i.e.

$$V[f] \;=\; V[g]\,, \qquad \text{for any} \quad f, g \in \omega(c_0)\,. \tag{4.3.24}$$

In particular

$$V[f(t,\cdot)](t) = V[f_0(\cdot)](0) \quad \text{for all } t \ge 0, \quad \text{and any } f_0 \in \omega(c_0)\,. \tag{4.3.25}$$

Consequently, the special form of

$$\frac{d}{dt}V[c(t,\cdot)] = -\frac{1}{2}\int_0^\infty\!\!\int_0^\infty (\ln(K\;c(t,u)c(t,w)) - \ln(F\;c(t,u+w))) \times (K\;c(t,u)c(t,w) - F\;c(t,u+w))\,du\,dw \tag{4.3.26}$$

is of use, implying that

$$K\;f(t,u)f(t,w) \;=\; F\;f(t,u+w), \qquad \text{for all} \quad t \ge 0\,. \tag{4.3.27}$$

Step 6. On account of Eq. (4.3.27), Eq. (4.3.16) itself may now easily be used to see that all solutions $f : t \mapsto f(t, \cdot)$ satisfying initial data $f_0 \in \omega(c_0)$ are equilibria.

Finally, uniqueness of equilibria proves that $\omega(c_0) = \{\bar{c}\}$ where $\bar{c}$ is defined in (4.3.19); and (3.b) above gives: $c \rightharpoonup \bar{c}$.

□

4.4 Critical Analysis and Perspectives

Condensation fragmentation phenomena can be of great interest in several fields of applied sciences. For instance, phenomena of the kind seen above may have to be modeled in the fluid dynamics of multiphase flow. Indeed, several interesting industrial problems are such that a primary flow carries coagulating or fragmenting materials, say clusters or bubbles. It follows that there are strong motivations to model not only the fundamental phenomena dealt with in this chapter, but also phase transition phenomena in multiphase flow of particles dispersed in a continuum stream.

Kinetic models can be useful in the following cases: either when the mean free path of the dispersed particles is large compared with that of the primary flow particles, or when the size of the dispersed particles is sufficiently small, for instance of the same order of magnitude as those of the primary flow.

The interest into condensation fragmentation phenomena is not limited to fluid dynamics. Interesting phenomena occur in biology as documented in [AZa]. This aspect will be further discussed in Chapter 6, where the phenomenon of aggregation and condensation of cells in a vertebrate is considered in connection with formation of solid tumors.

However, despite the above motivations, the modelling still needs to be developed towards physically acceptable descriptions of real phenomena. This last section concerns a critical analysis of the class of models reviewed in this chapter, and an indication of research perspectives.

What follows does not cover the whole variety of open problems, but simply represent a selection of them following authors' bias. Some of these problems will be partially dealt with in Chapter 9 in a mathematical framework more adherent to the Boltzmann equation. Bearing this in mind, the following topics, among several others, may be cited as research perspectives.

• The above evolution model is homogeneous in space. This assumption is certainly a limit of the model applicability to the description of real flow conditions. Models with diffusion have been dealt with by various authors mainly in connection with the qualitative analysis of the evolution problems. In particular, and among others, discrete type models are dealt with by Laurencot [LAa] and Wrzosek [WRa]; however, the same method can be developed for continuous models. In both cases, on the one hand the space dependent mathematical problem is certainly interesting and the qualitative analysis valuable; on the other hand, diffusion models are valid only for small diffusivity, and applications need models able to describe strong non equilibrium conditions. More recently Lachowicz and Wrzosek [LCa], have proposed a model with nonlocal interaction very closed to the one by Jager and Segel (we have cited in Chapter 3) and have provided global; existence proofs.

• The above model is based on the assumption that the physical system is constituted by one fluid only, namely the coagulating fragmenting fluid. Several interesting physical systems exist such that the clusters are conveniently introduced to model a second fluid immersed (mixed) into the first fluid. In this case it is necessary to analyse not only the mixture evolution, but also the interactions between the first and the second fluid. Modelling requires, in addition, an analysis of the order of magnitude of the mean free path of the first fluid particles to be compared with the mean size of the clusters. According to the result of the comparison, the best choice may be made concerning the model: in the framework of kinetic theory when the ratio is of order one; in the framework of the continuum mechanics when the ratio is small compared with unity.

• Identification of the terms F and K is based on purely phenomenologic arguments and makes no reference to a microscopic analysis related to molecular properties. It follows that the model may provide information only on the qualitative behavior of the system evolution, and some microscopic modelling needs to be constructed in order that a quantitative analysis, related to the properties of the coagulating fragmenting fluid material, may be developed. An interesting alternative to continuous models is the use of discrete models; i.e. those such that the size of the clusters is discrete. This assumption may help the identification of some interesting physical connections of the model. A simple example is that studied by Kreer and Penrose [KRa], which reads

$$\frac{dc_i}{dt} = \sum_{k=1}^{i-1} a_{(i-k),k} c_{(i-k)} c_k - 2 \sum_{k=1}^{\infty} p_{i,k} c_i c_k \, , \tag{4.4.1}$$

and describes a process where two clusters of sizes k and i can coagulate to a cluster of size $i + k$ with probability $p_{i,k} = p_{k,i} > 0$. The model is a very simple one, in that there is no fragmentation of clusters. Indeed, one can generalize the various models we have seen in the preceding sections to the discrete case. Some difficulties reside in the identification of the transition probability. On the other hand, in the discrete case interesting physical properties may be proved even for an oversimplified model such as that represented in Eq. (4.4.1). For instance, the so-called ***dynamical scaling*** assumption is proved in [KRa]. This conjecture asserts that, after a sufficiently long time, the large clusters distribution becomes independent of the initial value, and approaches a distribution which, on suitable time and length scales, has a self-similar profile

$$c_i \cong t_a c^*(i\, t^{-b}) \, , \qquad a, b \in \mathbb{R},$$

where c^* is the so-called dynamical scaling function.

• The above model is based on the assumption that clusters may change state only because of collisions with other clusters. One may assume, on the contrary, that the time evolution of the clusters size does not, or not only, depend on the collision mechanics. For instance vaporization phenomena can occur in non equilibrium conditions. As before, a microscopic analysis related to molecular properties of the material may provide useful information on this type of modelling.

• Models with multiple fragmentation have been proposed by McLaughlin, Lamb, and McBride, see [MCa] and the bibliography therein cited. The approach is again phenomenological, as it is for models with simple fragmentation. Further development of this kind of modelling, with a special attention to the case with collisions, may lead to results of great interest for the applications, and in particular to the description of sprays.

4.5 References

[AIa] Aizenman M. and Bak T.A., Convergence to equilibrium in a system of reacting polymer,*Commun. Math. Phys.*, **65** (1979), 203–230.

[ASa] Ash R. B., **Real Analysis and Probability**, Academic Press, New York, (1970).

[AZa] Ackleh A.S., Fitzpatrick B.G. and Hallam T.G., Approximation and parameter estimation problems for algal aggregation models, *Math. Models Meth. Appl. Sci.*, **4** (1994), 291–311.

[BAa] Ball J.M., Carr J. and Penrose O., The Becker-Döring cluster equations: basic properties and asymptotic behaviour of solutions, *Comm. Math. Phys.*, **104** (1986), 657–692.

[DUa] DUBOVSKII P.B. and STEWART I.W., The order of singularity for the sationary coagulation equation, *Appl. Math. Lett.*, **8** (1995), 17–20.

[EDa] EDWARDS R.E., **Functional Analysis: Theory and Applications**, Holt, Rinehart and Winston, New York, (1965).

[ERa] ERNST M.H., ZIFF R.M., and HENDRIKS E.M. Coagulation processes with a phase transition, *J. Coll. Interf. Sci.*, **97** (1984), 266–277.

[KRa] KREER M. and PENROSE O., Proof of dynamical scaling in Smoluchowski's coagulation equation with constant kernel, *J. Stat. Phys.*, **75** (1994), 389–407.

[LAa] LAURENCOT P. and WRZOSEK D., The Becker-Döring model with diffusion: Basic properties of solutions, *Colloq. Math.*, **75** (1998), 245–269.

[LCa] LACHOWICZ M. and WRZOSEK D., A nonlocal coagulation fragmentation model, preprint Inst. Appl. Math. Mech. Univ. Warsaw n. 49, January, (1999).

[MCa] MCLAUGHLIN D., LAMB W., and MCBRIDE A., An existence and uniqueness results for a coagulation and multiple fragmentation equation, *SIAM J. Math. Anal.*, **28** (1997), 1173–1190.

[MCb] MCLAUGHLIN D., LAMB W., and MCBRIDE A., A semigroup approach to fragmentation models, *SIAM J. Math. Anal.*, **28** (1997), 1158–1172.

[STa] STEWART I.W., A global existence theorem for the general coagulation fragmentation equation with unbounded kernels, *Math. Proc. Camb. Phil. Soc.*, **107** (1990), 573–578.

[STb] STEWART I.W., A uniqueness theorem for the coagulation fragmentation equation, *J. Appl. Math. Phys.*, **41** (1990), 917–924.

[STc] STEWART I.W., On the coagulation fragmentation equation, *J. Appl. Math. Phys.*, **41** (1990), 917–924.

[STd] STEWART I.W., Density conservation for a coagulation equation, *J. Appl. Math. Phys.*, **42** (1991), 746–756.

[STe] STEWART I.W. and DUBOVSKII P.B., Approach to equilibrium for the coagulation fragmentation equation via a Lyapunov functional, *Math. Meth. Appl. Sci.*, **19** (1996), 171–185.

[WHa] WHITE W.H., Global existence theorem for Smoluchowski's coagulation equation, *Proc. Amer. Math. Soc.*, **80** (1980), 273–276.

[WRa] WRZOSEK D., Existence of solutions for the discrete coagulation fragmentation model with diffusion, *Topological Methods in Nonlinear Analysis*, **9** (1997), 279–296.

Chapter 5

Kinetic Cellular Models in the Immune System Competition

5.1 Kinetic Models Towards Immunology

Chapters 5 and 6 give some generalizations of the Boltzmann equation in the field of immuno-mathematical theories, an application field of generalized kinetic models which is nowadays of great relevance, and that may open extremely interesting research perspectives for applied mathematicians.

Before entering into the modelling details, and related mathematical problems, a concise discussion on the possibilities of immuno-mathematical theories is convenient. The argument concerns both the present chapter and the next one, and that may give hints on a new and certainly fascinating research field.

Common denominator of any physico-mathematical theory is the effort to provide detailed description and insight of a physical system by means of some set of evolution (and equilibrium) equations characterized by directly measurable parameters. Even if this procedure is, in general, unable to provide a complete understanding of the physical system, it is suitable to acquire at least some information on the real system behavior.

However, in order for the theory to be reliable, it is clear that actual experimental observations must be used as its starting point. Only afterwards, qualitative and computational analysis of the problems generated by the theory and linked to the various observations of the system, may bring to its validation and possible developments.

Immuno-mathematical theories, in particular, are made to understand and describe aspects of the immune system behavior, and of its competition against aggressive hosts, for instance invasive tumor cells. In pursuing this objective one should be aware that the interaction between immunology and mathematical sciences is still extremely weak, almost non existing, especially if compared with that between mathematics and physics. Physics can rely on a large background of deterministic theories (or models) strongly supported by well-known interactions laws that rule the behavior of the various elements that constitute the physical system. Conversely, extremely few background fields are available in immunology, wherein determinism is almost totally lost.

Actually, although the relevant features, in common to all individuals, of several of the environmental phenomena of a vertebrate may be properly defined and identified, their quantitative aspects often undergo deep changes from case to case. Consequently, and almost unavoidably, one is led to introduce stochastic methods and analysis. The scale to reach determinism is indeed too small to allow efficient observations.

Moreover, the real problem may be too difficult to justify, or suggest, waiting for further developments in the experimental research field. Hence a possible conclusion is that proposed by Wigner [WIa] about three decades ago. He gave the statement of ***unreasonable effectiveness of mathematics in natural sciences***. As a matter of fact even nowadays, despite the great efforts devoted to this matter, and despite the vast mathematical biology literature which consists of several specialized journals and books, biological sciences take very little advantage from the collaboration with mathematics.

An interesting contribution to this discussion may be found in a nontraditional paper by Curti and Longo [CUa], two highly experienced immunologists. Their survey, among various ideas, calls for scientists attention on the following aspects of the general problem:

a) Modelling should be developed at the cellular level; all phenomena that are relevant for immunology develop at such a scale.

b) Determinism is lost in cellular interactions; only stochastic behaviors may be meaningfully considered in the model.

c) The dialogue between immunologists and applied mathematicians is made difficult by the deeply different research styles and languages.

d) Despite the great advances and successes reached in molecular biology, the fall over on therapies, for instance therapies against cancer, has been very modest.

The contents of Chapters 5 and 6 is somehow consistent with the above four items. In particular, according to items a)–b), modelling is developed at the cellular level, cell interactions are dealt with in a stochastic framework, and fluctuations are included without hiding the basic phenomena.

Chapter 5 is divided into five sections.

Section 5.1 is this introduction.

Section 5.2 describes a possible mathematical strategy to develop immuno-mathematical theory related to a vertebrate physical system. Particular attention is given to identify the scales that characterize the phenomenon.

Section 5.3 provides the modelling of a phenomenological system.

Section 5.4 develops the derivation of a suitable evolution equation for an invasive tumor cell population versus the immuno system cell population.

Section 5.5 deals with a review on the mathematical results about qualitative analysis of initial value problems. It also contains a review of the simulations related to the application of the model, and a discussion on research perspectives.

The model's background resides both on kinetic theory of gases and on ***mathematical biology.*** The interested reader is then addressed to the book by Murray [MUa]. Particular attention has to be given to the papers about ***modelling cellular phenomena.*** Indeed, cellular activities play a crucial role in the physical systems we are dealing with, and a mathematical interpretation of cellular dynamics can effectively contribute to a correct modelling of the interactions between tumors and immune system. The book by Segel [SEa] is an excellent guide to mathematicians interested in these phenomena. General aspects of ***modelling tumor dynamics*** are also of importance; they are dealt with in the book by Den Otter and Ruitenberg [DEa] devoted to the analysis of angiogenic factors, in the collection of papers edited by Green, Cohen and McCluskey [GRa] about the mechanisms of tumor immunity, and in the essay by Nossal [NOa] on general aspects of the immune system and its competition with tumor aggressions. Interpretation and simulation of experimental results is proposed in the book by Asachenkov et al. [ASa]. A more specialized literature can be recovered in the review paper [BEf] which also indicates both recent developments and research perspectives.

5.2 Scaling in Kinetic Cellular Models

As already mentioned, in this chapter we present some generalized kinetic models and basics on the so-called (cellular) kinetic theory for tumor–immune system competition. The class of models reported in this chapter can be regarded as a particular development and application of population dynamics modelling with stochastic interactions described in Chapter 3.

The evolution of a cell, as described in the specialized literature [ABa], is ruled by genes contained in its nucleus. Genes are activated, or suppressed, by signals detected by receptors localized on the cell surface and then transducted to its nucleus. Receptibility of particular signals can modify the usual behavior of a cell. In extreme situations, particular signals may induce a cell to reproduce in the form of identical descendants, this is the so-called ***clonal expansion***, or to die and disappear without apparent trace, this is the so-called ***apoptosis*** or ***programmed death***. In the case of clonal expansion a competition is activated between tumor cells and cells of the immune system. If immune system cells are active and able to recognize tumor cells, then they may develop a reactive mechanism that can even destroy the aggressive host. On the contrary, if the immune system is not sufficiently active, the tumor may progressively grow. Activation and disactivation of immune cells is influenced by cytokine signals.

This simple description clearly indicates that tumors evolution, and their competition with the immune system, has to be modelled at a cellular level. This point of view is well recognized in the literature, as documented in the paper by Tomlison and Bodmer [TOa], where simple phenomenological models are proposed to simulate, in a deterministic framework, the effects of programmed cell death and of cells differentiation. These two events are both recognized to be responsible for tumor cells proliferation.

Recently [BEa] and [BEb] a new methodological approach has been proposed to describe the interaction between tumors and immune system. Modelling is developed in a mathematical framework typical of kinetic theory of gases. Starting point of the model is the description of cellular interactions, to be regarded as the microscopic level. Then, with a procedure similar to that used in phenomenologic kinetic theory, an evolution equation is derived for each population distribution function. Several interacting cellular populations are considered, respectively pertaining to the tumor, the host environment and the immune system.

Characteristic feature of the model is that each cell is attributed with a certain physical state. Interactions can either induce a cell to change its state, or cause cell proliferation and destruction. The system evolution is obtained by suitable balance equations, that rule the time evolution of the populations number density functions.

This class of models refers to the tumor early growth in a regime that was defined the ***free cells regime***, meaning that tumor cells are not yet condensed in a macroscopically observable quasi-spherical tumor. This stage is of particular importance because the competition between tumors and immune system can still be addressed toward tumors depletion. Indeed, on free cells regime appropriate external actions may still produce a reliable enforcement of the immune system.

A crucial role in activating immune defense is played at this stage by cytokine signals. Cellular interactions are ruled [FOa] by signals that the various cells emit and receive through a complex image recognition process. Thus, interactions apparently developed at the cellular level are in fact ruled by mechanisms that live at a lower scale.

After a suitable maturation time, tumor cells start to condense and aggregate into a quasi-spherical nucleus. An extended tumor interacts with the outer ambient, say environmental and immune system cells, only through its surface and within a layer where angiogenesis phenomena take place. Here, cellular phenomena overlap with typical macroscopic behaviors such as diffusion or, more in general, phenomena that can be related to conservation and evolution of variables of macroscopic character.

In conclusion, it has to be made clear that a reliable immuno-mathematical theory needs to include descriptions at all scales, which we schematize as follows:

- ***sub-cellular scale***, which refers to the emission and reception of signals that are finalized to the regulation of cellular activities.
- ***cellular scale***, which refers to the main activities of the cells, such as loss of differentiation, proliferation of tumor cells, competition of immune cells, and so on.
- ***macroscopic scale*** which refers to phenomena which are typical of continuum systems: convection, diffusion and so on.

The analysis developed in what follows mainly refers to the free cellular regime. The lower scale is analyzed only to model cellular behaviors; the higher scale only when related to the system dynamics at cellular scale. At this scale, the following activities will be modelled among others:

- ***clonal expansion*** of tumor cells;
- ***feeding activity*** from environmental cells to tumor cells;
- ***inhibition activity*** from tumor cells to immune cells;
- ***cooperation activity*** from inhibited immune cells to tumor cells;
- ***weaking and destruction activity*** from active immune cells to tumor cells.

Sequential steps to obtain the evolution model are the following:

Step 1. *Selection of the cell populations which play a role in the system qualitative and quantitative evolution.*

Step 2. *Identification of the cells activity, that labels each cell of the same population, and of the cytokine signals that rule such an activity.*

Step 3. *Construction of the model that simulates the dynamics of cell-cell interactions.*

Step 4. *Construction of the model that simulates the influence of cellular dynamics at the sub-cellular scale, and its feed-back on the dynamics of cells-cell interactions.*

Step 5. *Derivation of suitable evolution equations that may describe cellular number densities as functions of time and of cells activity.*

Step 6. *Derivation of suitable evolution equations that may describe the system macroscopic observables in a fashion to be consistent with the observed cellular behaviors.*

As already discussed, and similarly to the classical kinetic theory, it is necessary to identify the cellular regimes that characterize the system behavior. Our modelling refers to ***free (dispersed) cells regime***, meaning that the cells move freely and homogeneously in space, and that interactions are restricted to binary encounters. The opposite picture is the ***condensed (packed) cells regime***, a physical situation wherein the tumor cells are already packed in a quasi-spherical volume. In this case, the immune system cells and the feeding environmental cells interact among themselves as in the free cells regime, but with interactions that are restricted to the tumor surface. Indeed, distinction between these two regimes is not so easy, and intermediate regimes may also be taken into account. Nevertheless, early neoplastic growth refers to the free cells regime, and the models proposed in this chapter are devoted to it.

5.3 Phenomenological System and Modelling

A vertebrate is a body made up by a variety of interacting cells. In particular, in our present modelling we consider tumor cells, environmental cells, and immune-system cells. Each cell is tagged by a variable, here in after called the ***state***, either scalar or multi-dimensional, which functionally defines the cell activities.

Kinetic modelling is based upon describing the statistical behavior of each population, and upon deriving suitable evolution equations for the corresponding distribution functions. First step toward the modelling is a phenomenological observation of the system and, in particular, defining the interacting cell populations. Then, for each population, the set has to be identified with those cellular activities that are seen to be the most relevant ones for that population.

The set of peculiar cellular activities generally differs from population to population. Each activity is tagged by a dimensionless variable that refers to the strength with which the activity is performed. For the sake of model simplicity, it will here be assumed that the state is a scalar variable, say u, comprehensive of all the cell activities, with values in the range $[-1, 1]$. When $u = 1$ the cell performs its peculiar activities at the highest possible degree. Negative values correspond to a complete inhibition of the cell's characteristic functional behavior, and even to the onset of new features that contrast with the positive ones and may produce opposite results. The strongest negative behavior occurs for $u = -1$.

In other words, it is assumed that the peculiar activities of each cell of a certain population are spread over a certain range, whereon each cell may choose its "positive" or "negative" behavior. Negative activities even counteract positive ones. In this way, the model is possible both of the dormant and the proliferative tumor behavior on one side, and both of the tumor inhibitory and tumor stimulatory action by the immune and environmental cells on the other side.

The switch between the opposite effects of an activity is particularly evident in the case of immune system cells. Indeed, starting from their inhibitory action of tumor's aggressivity, they may degenerate up to a cooperative action. Similarly, during long periods of their natural history, tumor cells may assume a dormant behavior (u approaching $u = 0$), or even be pharmacologically induced to undergo programmed cell death (apoptosis corresponding to $u = -1$). The goal, at least in principle, is to understand how to favor the trend of tumor cells distribution toward the range of negative values of the variable u.

Bearing all this in mind, the cell populations and tumor-host relationships are identified by means of the following assumptions:

Assumption 5.3.1. *Cells that actually play the game may be subdivided into $n = 3$ main populations:* ***tumor cells****, characterized by anomalous proliferative activity and by poor receptibility to inhibitory and apoptotic signals;* ***environmental cells****, characterized by both promoting and inhibiting tumor cells;* ***immune system cells****, potentially able to either strongly hamper or even favor tumor growth. The activities of all these cell populations are regulated by soluble cytokine signals and cell membrane signals.*

Assumption 5.3.2. *The functional* ***state*** *of each cell of a given population is assigned by a real number u taking values in the interval $[-1, 1]$. Positive values of u denote:* ***high proliferation*** *for tumor* *cells;* ***adequate energy supply*** *for the environment;* ***anti-tumor activity*** *for immune system cells;* ***activation*** *of the immune and environmental cells by cytokine and stimulatory signals. Negative values correspond to opposite activities:* ***dormancy*** *of tumor cells;* ***ischemic necrosis*** *due to poor blood supply for the environment;* ***tumor stimulating activity*** *for the immune cells;* ***inhibition*** *of the immune system by cytokine signals.*

Assumption 5.3.3. *The statistical distribution of the system populations is described by a set of number density functions*

$$\mathbf{f} = \{f_i\}_{i=1}^n; \qquad f_i : (t,u) \in [0,T] \times [-1,1] \mapsto f_i(t,u) \in \mathbf{R}_+ \,, \tag{5.3.1}$$

where $f_i(t,u)\,du$ defines the number per unit volume of cells of the i-th population that at time t are in a state within the interval du centered at u. The number of all possible interacting populations will generically be denoted by n although the calculations that follow are developed in the case $n = 3$.

Admittedly, defining the cell populations that are involved in this lethal game, and defining their functions, implies gross simplifications. The difficult target consists in sketching and summarizing the critical features that shape the process. In a sense, some advantages may be found by considering that the observed reality is the result of an accumulation of gene mutations that cause tumor cells unrestrained proliferation and their lack of response to the environment signals. Accumulation of gene mutations and intense proliferative activity make tumor cells different from the normal cells of the same tissue. Thus, each tumor may be characterized by a particular set of gene mutations and by a more or less rapid expandability and independence from those environmental signals that regulate cells proliferation or apoptotic death. Detailed descriptions of this physical phenomena are reported in the surveys published in [ADa].

Modelling cell interactions will be developed according to the following hypotheses:

Assumption 5.3.4. *Only binary interactions are considered. They are divided into* ***mass conservative*** *encounters and interactions that produce* ***proliferation*** *and* ***destruction*** *of cells. Mass conservative encounters are characterized by: the* ***encounter rate*** η *and the* ***transition probability density*** ψ*. Proliferative encounters are described by: the* ***proliferation rate*** p *and the* ***proliferation probability density*** φ*. Destructive encounters are described by: the* ***destruction rate*** d*.*

Assumption 5.3.5. ***Conservative encounters*** *occur between pairs of cells of the same or different populations, and produce a change of state of one of the cells involved. These interactions are quantitatively described by the* ***conservative transition rate***

$$\mu_{ij}(v, w; u) = \eta_{ij}(v, w)\psi_{ij}(v, w; u)\,, \tag{5.3.2}$$

which denotes the number density of pairwise encounters, per unit volume and unit time, between cells of the i*-th population in the state* v *and cells of the* j*-th population in the state* w*, that are of conservative nature because the* i*-cell (survives and) undergoes a transition from state* v *to state* u*.*

The ***conservative encounter rate*** $\eta_{ij}(v, w)$ is a number section of conservative pairwise encounters in the sense that

$$\eta_{ij}(v, w)f_i(t, v)f_j(t, w)dv\,dw$$

gives the number of conservative pairwise encounters, per unit volume and unit time, between cells of the i-th and j-th population in states within dv and dw at v and w respectively.

The ***transition probability density*** $\psi_{ij}(v, w; u)$ measures the probability that, when a (conservative) encounter takes place between a cell of population i in state v and a cell of population j in state w, the i-cell undergoes a transition from its initial state v to the final state u.

Assumption 5.3.6. ***Proliferative encounters*** *occur between cell pairs of the same or different populations, and generate proliferation of cells of one or both populations. These interactions are quantitatively described by the* ***proliferation rate***

$$\nu_{ij}(v, w; u) = p_{ij}(v, w)\varphi_{ij}(v, w; u) \tag{5.3.3}$$

which denotes the number density of pairwise encounters, per unit volume and unit time, between cells of the i-th population in the state v and cells of the j-th population in the state w, that are of proliferative nature because a new cell of population i is produced in the state u.

The ***proliferative encounter rate*** $p_{ij}(v, w)$ is a number section of proliferative pairwise encounters in the sense that

$$p_{ij}(v, w) f_i(t, v) f_j(t, w) dv\, dw$$

gives the number of pairwise encounters, per unit volume and unit time, between cells of the i-th and j-th population in states within dv and dw at v and w respectively, and that are of proliferative nature because a new cell of the i-th population is produced.

The ***proliferation probability density*** $\varphi_{ij}(v, w; u)$ measures the probability that, when a (proliferative) encounter takes place between a cell of population i in the state v and a cell of population j in the state w, a new cell of the i-th population is produced in the state u.

Assumption 5.3.7. ***Destructive encounters*** *occur between cell pairs of different populations, and produce the destruction of one or both cells. These interactions are quantitatively described by the* ***destruction rate*** *$d_{ij}(u, w)$ which denotes the number section of pairwise encounters, per unit volume and unit time, between cells of the i-th and j-th population in the states v and w respectively and which are of destructive nature because the i-th cell is destroyed.*

Terms φ and ψ are (conditional) probability densities with respect to the state variable u:

$$\int_{-1}^{1} \varphi_{ij}(v, w; u)\, du = \int_{-1}^{1} \psi_{ij}(v, w; u)\, du = 1\,, \tag{5.3.4}$$

for $i, j = 1 = 1, \ldots, n$ and $v, w \in [-1, +1]$.

Assumptions (5.3.4)–(5.3.7) define a microscopic behavior at the cellular level. They provide the necessary background to derive the evolution model.

5.4 Kinetic Evolution Equations

Evolution equations for densities f_i, where $i = 1, \ldots, n$, are obtained by means of suitable balances that equate the total time derivative of f_i to the difference between the conservative ***gain*** term G_i minus the conservative ***loss*** term L_i, plus the difference between the ***proliferation*** term P_i minus the ***destruction*** term D_i. Both the terms G_i and L_i are related to pairwise cell interactions with consequent i-th cell survival and change of state.

The set of equations is schematized by

$$\frac{\partial f_i}{\partial t} = G_i[\mathbf{f}] - L_i[\mathbf{f}] + P_i[\mathbf{f}] - D_i[\mathbf{f}]\,, \tag{5.4.1}$$

for $i = 1, \ldots, n$. All G_i, L_i, P_i, and D_i operate on $\mathbf{f} := (f_1, \ldots, f_n)$. This allows to take into account the possible relationships between different cell-types. In particular:

- G_i is obtained by summing all transitions, ruled by η and ψ, of i-th cells from any states v into the state u. They may occur when a cell of population i in any states v encounters a cell of any populations j in any states w, thus a summation over j is required together with double integration over the variables v and w.

- L_i is obtained by summing all transitions, ruled by η, of i-cells from the state u to any other state. They may occur when a cell of population i in the state u encounters any j-cells in any states v, thus a summation over j is required together with integration over v.
- P_i is obtained by summing all proliferations, ruled by p and φ, of i-th cells into the state u. They may occur when a cell of population i in any states v encounters any j-th cells in any states w, thus a summation over j is also needed together with integration over v and w respectively.
- D_i is obtained by summing all destructions, ruled by d, of i-th cells from the state u. They may occur when a cell of population i in the state u encounters any j-th cells in any states v, which requires summation over j and integration over v.

On account of the above detailed list, the evolution equations may be written as follows, for $i = 1, \ldots, n$,

$$\begin{aligned}
\frac{\partial f_i}{\partial t}(t,u) &= \sum_{j=1}^{n} \int_{-1}^{1} \int_{-1}^{1} \eta_{ij}(v,w)\psi_{ij}(v,w;u) f_i(t,v) f_j(t,w)\, dv\, dw \\
&\quad - f_i(t,u) \sum_{j=1}^{n} \int_{-1}^{1} \eta_{ij}(u,v) f_j(t,v)\, dv \\
&\quad + \sum_{j=1}^{n} \int_{-1}^{1} \int_{-1}^{1} p_{ij}(v,w)\varphi_{ij}(v,w;u) f_i(t,v) f_j(t,w)\, dv\, dw \\
&\quad - f_i(t,u) \sum_{j=1}^{n} \int_{-1}^{1} d_{ij}(u,v) f_j(t,v)\, dv\,. \qquad (5.4.2)
\end{aligned}$$

It is plain that the model reported in Eq. (5.4.2) is a general one and can be regarded only as a starting framework for specifically applicable models. Special applications involve detailed identification of the function matrices

$$[\eta] = [\eta_{ij}], \quad [p] = [p_{ij}], \quad [d] = [d_{ij}], \quad [\psi] = [\psi_{ij}], \quad [\varphi] = [\varphi_{ij}].$$

On the other hand, this identification may require further assumptions on the analytic structure of some of the above mentioned terms. For this purpose, the idea commonly followed is that the simpler the structures the more convenient they are to design possible experiments addressed to realize the above identification. In principle, identification should be based upon a detailed analysis of the cellular behavior properties and of their characterization at sub-cellular level. Unfortunately, this topic is only partly understood. It is in fact a research perspective, and the present knowledge is based only upon a purely phenomenological identification.

The procedure proposed in [BEc] is based on some particular assumptions on the analytic structure of the above matrix entries, so that identification may be reduced to a limited number of constants. For instance, for each choice of the parameters v, w, the (corresponding) densities $u \mapsto \psi(v, w; u)$ and $u \mapsto \varphi(v, w; u)$ are assumed to be continuous over the whole u range $[-1, 1]$. In addition, the dependence on the parameters v and w is assumed to be achieved only through the output state mean value $m = m(v, w)$ and its standard deviation $\sigma = \sigma(v, w)$. Moreover, the standard deviation is presumed to be sufficiently small, and the probability densities ψ and φ sufficiently similar to delta functions.

The description reported in what follows mainly refers to [BEc] and [BEd], where also specific simulations are reported. In the above papers, simulation results are presented, based on such identification, that show an interesting description of the system behavior. Of particular relevance is the fact that this model may successfully describe both the two conflicting situations: either fast growth of immune system and depletion of tumor cells, or fast growth of tumor cells and inhibition of immune system. Both behaviors are experimentally observed, and they are found to be connected by dormancy periods that may be surprisingly long. Indeed, in the framework of this theory, dormancy time intervals correspond to slow activation of tumor cells while changing from negative to positive values of their activation state.

To be specific, we consider that the physical system may be represented by Eqs. (5.4.2) in the case of three interacting populations: tumor cells, immune system cells, and environmental cells, respectively corresponding to $i = 1, 2, 3$. As a further simplification, we assume that the number density of environmental cells remains a constant during the observation time of the phenomenon. Such a crude simplification, though compressing the immune system into only one population, it is sufficient to draw a careful description of the relevant features of the system behavior.

The cellular interactions that are considered to be significant for the system evolution are:

- conservative encounters between tumor and immune cells,
- feeding encounters between tumor and environment cells, that may generate tumor growth,
- cooperative encounters between degenerate immune system and tumor cells,
- destructive interactions between active immune system and tumor cells, that may generate tumor depletion,
- proliferative interactions between immune and tumor cells, that may generate proliferation of immune cells,
- cytokine actions which affect the output of conservative interactions.

As it is seen, only a limited angle of a vast panorama is taken into account. In particular, the last item states that only conservative interactions may be modified by cytokine signals. Conversely, destructive and proliferative encounters are completely ruled by the values of the cellular activation states, although attained through the mediation of the above signals. The following assumptions provide identifications of the above matrices.

Concerning ***conservative encounters***, the following assumptions are stated:

Assumption 5.4.1. *Environmental temperature remains constant during the whole observation time, and hence the encounter rates* η_{ij} *remain constant as well. Interactions between tumors and immune system cells are the only ones that may generate a change of state of any of the cells involved (nontrivial interactions); they are such that* $\eta_{12} = \eta_{21} = \alpha$, where $\alpha \geq 0$ *is a constant. All other encounters do not generate state changes:*

$$\psi_{11} = \psi_{22} = \psi_{33} = \psi_{13} = \psi_{31} = \psi_{23} = \psi_{32} = \delta(v - u);$$

therefore, their contributions to gain term $G_i[\mathbf{f}]$ *and loss term* $L_i[\mathbf{f}]$ *are equal and cancel out.*

Assumption 5.4.2. *The transition probability densities* $\psi_{ij}(v, w; u)$ *are subject to the two functional constraints implicitly stated by assigning: the expected value* $m_{ij} = m_{ij}(v, w)$, *and the variance* $\sigma_{ij} = \sigma_{ij}(v, w)$, *of the i-th cell output state, when it leaves the state* v *because of an encounter with a j-th cell in the state* w.

Assumption 5.4.3. *The expected value* m_{12} *for the output state* u *of a tumor cell in the state* v *after a conservative encounter with an immuno system cell in the state* w *is given by*

$$\begin{cases} v > 0\,,\; w \geq 0 & \Rightarrow \quad m_{12}(v, w) = v - \beta_{12}^{+} w v\,, \\ v > 0\,,\; w < 0 & \Rightarrow \quad m_{12}(v, w) = v - \beta_{12}^{-} w(1 - v)\,, \end{cases} \tag{5.4.3}$$

and

$$\begin{cases} v \leq 0\,,\; w \geq 0 & \Rightarrow \quad m_{12}(v, w) = v\,, \\ v \leq 0\,,\; w < 0 & \Rightarrow \quad m_{12}(v, w) = v - \beta_{12}^{-} w(1 - v)\,, \end{cases} \tag{5.4.4}$$

where $\beta_{12}^+, \beta_{12}^- \in [0,1]$.

Assumption 5.4.4. *The expected value m_{21} for the output state u of an immuno system cell in the state v after a conservative encounter with a tumor cell in the state w is given by*

$$\begin{cases} w \geq 0 \quad \Rightarrow \quad m_{21}(v,w) = v - \beta_{21} w(1+v)\,, \\ w < 0 \quad \Rightarrow \quad m_{21}(v,w) = v\,, \end{cases} \tag{5.4.5}$$

where $\beta_{21} \in [0,1]$.

Assumption 5.4.5. *The bone marrow constantly produces active immune cells, uniformly distributed over the positive activation state*

$$s_2(t,u) = \begin{cases} s_2 & \text{if } u \geq 0\,, \\ 0 & \text{if } u < 0\,. \end{cases} \tag{5.4.6}$$

Remark 5.4.1. *Assumption 5.4.3 corresponds to the following* ***phenomenologic*** behavior.

If both the interacting tumor and immune cells are active, the tumor cell state decreases, but not below the dormant state: $v = 0$. In this case: $0 \leq m_{12}(v,w) \leq v$.

If the tumor cell is active and the immune cell is degenerate, then the tumor cell state increases, but not above the highest activation state $v = 1$. In this case: $v \leq m_{12}(v,w) \leq 1$.

If the tumor cell already exhibits its maximum activity $v = 1$, this cannot increase any further: $m_{12}(v = 1, w < 0) = 1$. Moreover, whatever the state of the tumor cell is it does not change if the immune system cell is inactive: $m_{12}(v, w = 0) = v$.

If the tumor cell is dormant and the immune cell is active, then the tumor cell state does not change. If the tumor cell is dormant and the immune cell is degenerate, then the tumor cell state increases.

Remark 5.4.2. *Assumption 5.4.4 corresponds to the following phenomenologic behavior.*

Whatever the state of the immune cell is, if the interacting tumor cell is active then the immune cell state decreases, but not below the maximum degenerate state: $v = -1$.

The higher the tumor cell activity w, the stronger the immune cell disactivation. If the immune cell already exhibits its maximum degeneration state $v = -1$, this cannot decrease any further.

Remark 5.4.3. *For simplicity, the role of cytokine signals is identified by assuming that signals may only affect conservative encounters. This means that they only operate on the β parameters. The activity of cytokine signals depends on both the signal type and on the signal strength.*

Concerning now ***proliferative encounters*** and ***destructive encounters***, it is assumed that proliferations and destructions occur without changing the interacting cell state:

$$\varphi_{ij}(v, w; u) = \delta(v - u)\,, \qquad \text{for } i, j = 1, 2, 3\,.$$

Destructive and proliferative events are assumed to depend upon the activation state of the interacting pair cells on the basis of the following model:

Assumption 5.4.6. *Interactions between a tumor cell in the state v and, respectively, a tumor an environmental or an immune system cell in the state w, are ruled by the following terms of proliferative and destructive nature:*

$$b_{11}(v, w) := p_{11}(v, w) - d_{11}(v, w) = 0\,, \tag{5.4.7}$$

$$\begin{cases} v > 0\,,\ w \geq 0 \ \Rightarrow b_{12}(v, w) := p_{12}(v, w) - d_{12}(v, w) = -\delta_{12} w\,, \\ v > 0\,,\ w < 0 \ \Rightarrow b_{12}(v, w) := p_{12}(v, w) - d_{12}(v, w) = -\gamma_{12} v w\,, \\ \qquad\qquad v \leq 0 \ \Rightarrow b_{12}(v, w) := p_{12}(v, w) - d_{12}(v, w) = 0\,, \end{cases} \tag{5.4.8}$$

and

$$\begin{cases} v > 0 \Rightarrow b_{13}(v,w) := p_{13}(v,w) - d_{13}(v,w) = \gamma_{13} v w H(w)\,, \\ v \leq 0 \Rightarrow b_{13}(v,w) := p_{13}(v,w) - d_{13}(v,w) = 0\,, \end{cases} \tag{5.4.9}$$

where H denotes the Heaviside function: $H(u) = 1$ if $u \geq 0$ and $H(u) = 0$ if $u < 0$.

Assumption 5.4.7. *Interactions between an immune cell in the state v and an environmental or a tumor cell in the state w are ruled by the following terms of proliferative and destructive nature*

$$\begin{cases} w \geq 0 \Rightarrow b_{21}(v,w) := p_{21}(v,w) - d_{21}(v,w) = \gamma_{21} \\ w < 0 \Rightarrow b_{21}(v,w) := p_{21}(v,w) - d_{21}(v,w) = 0\,. \end{cases} \tag{5.4.10}$$

$$\begin{cases} b_{22}(v,w) := p_{22}(v,w) - d_{22}(v,w) = 0\,, \\ b_{23}(v,w) := p_{23}(v,w) - d_{23}(v,w) = -\delta_{23} w H(w)\,. \end{cases} \tag{5.4.11}$$

Remark 5.4.4. *The above assumptions correspond to the following phenomenologic behavior.*
Proliferation of tumor cells is significant only when tumors interact with degenerate immune cells or with environmental cells. Proliferation only occurs for aggressive states, $v > 0$, and is directly proportional to the aggressivity of the tumor v and, respectively, to the negative activation of degenerate immune cells $-w$ or to the activation $w > 0$ of environmental cells.
Destruction of aggressive tumor cells only occurs in interactions with active immune cells and is directly proportional to the immune cells activation state.
Proliferation of immune cells occurs in interactions with active tumor cells and is independent on the activation states of the pair. Furthermore active immune cells are constantly produced by the bone marrow.

Destruction of immune cells only occurs on encounters with environmental cells which naturally control their growth. This action is proportional to the activation of the environmental cells.

It is plain that the assumptions above can be improved by a deeper analysis of the system behavior. Still, it is now possible to obtain a detailed evolution equations by specializing Eqs.(5.4.2):

$$\frac{\partial f_1}{\partial t}(t,u) = \int_{-1}^{1}\int_{-1}^{1} \alpha\psi_{12}(u, m_{12}(v,w)) f_1(t,v) f_2(t,w)\, dv\, dw$$
$$-f_1(t,u)\left\{\int_{-1}^{1} \alpha f_2(t,v)\, dv + \left[\gamma_{12} u \int_{-1}^{0} v f_2(t,v)\, dv\right.\right.$$
$$\left.\left. +\delta_{12}\int_{0}^{1} v f_2(t,v)\, dv - \gamma_{13} u \int_{0}^{1} v f_{30}(v)\, dv\right] H(u)\right\}, \qquad (5.4.12)$$

$$\frac{\partial f_2}{\partial t}(t,u) = \int_{-1}^{1}\int_{-1}^{1} \alpha\psi_{21}(u, m_{21}(v,w)) f_2(t,v) f_1(t,w)\, dv\, dw$$
$$+f_2(t,u)\left[\gamma_{21}\int_{0}^{1} f_1(t,v)\, dv - \int_{-1}^{1} \alpha\, f_1(t,v)\, dv\right.$$
$$\left. -\delta_{23}\int_{0}^{1} v\, f_{30}(v)\, dv\right] + s_2 H(u)\,, \qquad (5.4.13)$$

where the terms $\psi_{ij}(v,w;u) = \psi_{ij}(u, m_{ij}(v,w), \sigma_{ij}(v,w))$ are modelled to indicate the above said constraints and assumptions, and where the last term of Eq. (5.4.13) simulates the inlet from bone marrow. The role of the parameters that characterize the model is assessed as follows:

- α is related to the speed of evolution of the system and may be included in the time-scale.
- β_{12}^{+} characterizes the active immune system ability to deactivate the aggressive tumor cells.
- β_{12}^{-} characterizes the degenerate immune system negative action to enforce aggressive tumor cells.

- β_{21} characterizes the tumor cells ability to inhibit the immune system.
- γ_{12} characterizes the tumor cells proliferative behavior on their encounter with the inhibited immune system.
- γ_{13} characterizes the tumor cells proliferative behavior on their encounter with the host environment.
- γ_{21} characterizes the immune cells proliferative behavior on their encounter with aggressive tumor cells.
- δ_{12} characterizes the tumor cells suppression on their encounter with the active immune system.
- δ_{23} characterizes the immune cells suppression on their encounter with the host environment, that is: their natural extinction.
- s_2 characterizes the source of active immune cells from the bone marrow.

Therefore, parameters β are related to transitions in conservative encounters, parameters γ to proliferative activity, parameters δ to destructive activity. Practical simulations showed that the variance σ plays a secondary role. It has to be remarked that such a specialization of the various parameters is useful for their identification. For instance, in [BEe] it is discussed how cytokine signals can be related to the parameters β.

5.5 Qualitative Analysis, Applications, and Perspectives

This final section concludes the chapter reporting on the qualitative analysis of the initial value problem and on simulations related to applications of the model. This matter is proposed in two subsections which follow this concise introduction. Moreover, since we want to point out both the interesting and weak aspects of the above model, the last subsection contains a discussion on the research perspectives that may be addressed to improve the modelling, hopefully toward developing an immuno mathematical theory.

5.5.1 Qualitative analysis

Consider the initial value problem

$$\begin{cases} \dfrac{\partial \mathbf{f}}{\partial t} = \mathbf{J}[\mathbf{f}]\,, \\ \mathbf{f}(0,\cdot) = \mathbf{f}_0(\cdot)\,, \end{cases} \tag{5.5.1}$$

where $\mathbf{f} = \{f_i\}_{i=1}^n$, and where $\mathbf{J} = \{J_i\}_{i=1}^n$ represents the right-hand side of Eq. (5.4.2). The analysis follows the usual procedure, (see also Section 3.3):

a) definition of a suitable Banach space;
b) statement of solution to the initial value problem;
c) proof of existence, uniqueness, regularity, asymptotic behavior of solutions.

Referring to the literature already published on this topic, the analysis is generally developed in the space $\mathbf{X} = \big(L_1([-1,1]);\mathbb{R}\big)^n$ of the n-tuples $\mathbf{h}$ of real functions $h_i : [-1,1] \to \mathbb{R}$, $i = 1,\ldots,n$, such that $h_i \in L_1([-1,1];\mathbb{R})$. The space $\mathbf{X}$ is endowed by the norm

$$\|\mathbf{h}\| = \sum_{i=1}^n \|h_i\| = \sum_{i=1}^n \int_{-1}^1 |h_i(u)|\,du\,. \tag{5.5.2}$$

Solutions are sought in the positive cone $\mathbf{X}^+$ of $\mathbf{X}$, that is the subset of $\mathbf{X}$ of the functions $\mathbf{h} = \{h_i\}_{i=1}^n$, such that $h_i \geq 0$, for $i = 1,\ldots,n$.

The analysis can be related to the mild form

$$\begin{aligned} f_i(t,u) =& f_{i0}(u) \exp\left(-\int_0^t B_i[\mathbf{f}](s,u)\,ds\right) \\ &+ \int_0^t Q_i[\mathbf{f}](s,u) \exp\left(-\int_s^t B_i[\mathbf{f}](r,u)\,dr\right) ds\,, \end{aligned} \tag{5.5.3}$$

for $i = 1, \ldots, n$, that corresponds, with usual meaning of notations, to Problem (5.5.1) written in the form

$$\frac{\partial f_i}{\partial t} + f_i B_i[\mathbf{f}] = Q_i[\mathbf{f}], \qquad i = 1, \ldots, n. \tag{5.5.4}$$

A set of functions $\mathbf{f} = \{f_i\}_{i=1}^n$ is said to be a ***solution of the initial value problem*** (5.5.1) when it maps $[0, T]$ into $\mathbf{X}^+$, it is continuously differentiable on $[0, T]$, and it satisfies the differential evolution equation (5.5.3) and the initial conditions $\mathbf{f}_0$. Mathematical results are given in [ARb]. The analysis is also interesting for the applications. In fact, the presence of proliferative and competitive terms either activates a blow up of the solutions, or produces their asymptotic trend to zero. Qualitative analysis proves that the model is in fact able to describe both these behaviors. Further developments concerning the regularity of the solutions and their asymptotic behavior can be recovered in [ARc]. Here, we report the main ideas of the proofs. The analysis can be developed under the following

Assumption 5.5.1. *There exist* $c_1, c_2 \in \mathbb{R}^+$ *such that for all* $i, j = 1, \ldots, n$, *and almost all* $u, v, w \in [-1, 1]$, *the following inequalities are satisfied:*

$$0 \leq \eta_{ij}(u, v) \leq c_1, \tag{5.5.5}$$

and

$$0 \leq g_{ij}(v, w; u) \leq c_2, \tag{5.5.6}$$

where

$$g_{ij}(v, w; u) = \eta_{ij}(v, w)\, \psi_{ij}(v, w; u) + p_{ij}(v, w)\varphi_{ij}(v, w; u).$$

The operator $\mathbf{J}$ is locally Lipschitz continuous in $\mathbf{X}$ (the Lipschitz constant depends on c_1 and c_2) and hence there exists in $\mathbf{X}$ a unique solution $\mathbf{f}$ of Problem (5.5.1) on a time interval $[0, T]$, where

$T > 0$ only depends on c_1, c_2 and $||\mathbf{f}_0||$. Moreover, both operators $\mathbf{G} = \{G_i\}_{i=1}^n$ and $\mathbf{L} = \{L_i\}_{i=1}^n$ are positive and monotone. Therefore, by standard arguments already used in the theory of the Boltzmann equation [ARa], one can see that the solution $\mathbf{f}$ is non-negative, provided that the initial datum is non-negative (we follow the convention that $\mathbf{f} \geq 0$ if $f_i \geq 0$ for every $i = 1, ..., n$). Therefore, local existence follows:

THEOREM 5.5.1. *Let Assumption 5.5.1 be satisfied; then, for every $\mathbf{f}_0 \geq 0$ in $\mathbf{X}$, there exists $T > 0$ and a unique solution $\mathbf{f} : [0, T] \to \mathbf{X}^+$ of Problem (5.5.1).*

In the general case, the solution cannot exist globally in time, due to possible unboundedness of the local solutions, e.g., [ARb]. However, in the totally conservative and totally destructive cases, the following a priori estimate holds

$$||\mathbf{f}(t)|| \leq ||\mathbf{f}_0||\,, \qquad \text{for} \quad t \geq 0\,, \tag{5.5.7}$$

which guarantees global existence. Positivity of the solutions is a technical consequence of formula (5.5.3).

Global existence can be proved by a detailed analysis of the properties of the gain term. A rather general result is given in the last chapter. However, in the class of models dealt with in this chapter the crucial problem consists in analyzing the asymptotic behavior of the solution depending on bifurcation parameters. A computational analysis of this type is developed in [BEe] with reference to the influence of cytokine signals. This topic will be discussed later on.

5.5.2 Simulation problems

Numerical simulations of the model consist in solving the initial value problem (5.5.1) regarded as an initial value problem for a one-parameter family of ordinary differential equations.

Simulation results available in the literature show that, with suitable modelling of the parameters, an interesting description of the system behavior may be given. In particular, the model is able to describe the two conflicting situations: blow up of immune system and depletion of tumor cells, or blow up of tumor cells and inhibition of immune system. Both behaviors can be observed after a dormancy period that may be surprisingly long. In the framework of this theory, the longer the dormancy time interval, the slower the tumor cells activation from negative to positive values. This aspect is put in evidence in [BEb] and [BEc].

Computational analysis can be developed with the same technique that was outlined in Chapter 3 for conservative models. However, now the problem has to be handled more carefully. Here the L_1-norm is not preserved, and hence one loses control of the normalization that, conversely, regulates conservative models. Nevertheless, the above problems can be treated by computational schemes with much less difficulty than those related to the original Boltzmann model reviewed in Chapter 7.

In particular, a solution can be obtained by use of a suitable discretization of the variable u, that is

$$I_u = \{u_1 = -1, \dots, u_h, \dots, u_m = 1\}, \tag{5.5.8}$$

together with an interpolation of the densities $f_i(t,u)$ over u

$$f_i(t,u) \cong f_i^m(t,u) := \sum_{h=1}^{m} \chi_h(u) f_{ih}^m(t), \qquad i = 1, \dots, n, \tag{5.5.9}$$

where $f_{ih}^m(t) = f_i(t,u_h)$ for $h = 1, \dots, m$, and where $\chi_h = \chi_h(u)$ denotes the generic element of a set of fundamental interpolants corresponding to the above collocation and characterized by the property $\chi_h(u_k) = \delta_{hk}$, where δ_{hk} is the Kronecker delta. As already mentioned, fundamental splines, Lagrange polynomials and Sinc functions satisfy such a property.

This allows to transform the integrodifferential system into a suitable set of $n \times m$ ordinary differential equations

$$\frac{df_{ih}^m}{dt}(t) = J_{ih}[\mathbf{f}](t) := \sum_{j=1}^{n} \sum_{p,q=1}^{m} w_{ijhpq} f_{ip}^m(t) f_{jq}^m(t)$$

$$- f_{ih}^m(t) \sum_{j=1}^{n} \sum_{q=1}^{m} w_{ijhq}^* f_{jq}^m(t) + s_{ih}(t)\,, \qquad (5.5.10)$$

where $i = 1, \ldots, n, \;\; h = 1, \ldots, m,$ and the terms

$$w_{ijhpq} = \int_{-1}^{1} \int_{-1}^{1} \Big[\eta_{ij}(v, w) \psi_{ij}(v, w; u_h)$$

$$+ p_{ij}(v, w) \varphi_{ij}(v, w; u_h) \Big] \chi_p(v) \chi_q(w)\, dv\, dw \qquad (5.5.11)$$

and

$$w_{ijhq}^* = \int_{-1}^{1} \left[\eta_{ij}(u_h, v) + d_{ij}(u_h, v) \right] \chi_q(v)\, dv \qquad (5.5.12)$$

can be computed once and for all before starting the computations about the actual evolution of the system.

Therefore, the problem may be reformulated in terms of a system of nonlinear ordinary differential equations

$$\begin{cases} \dfrac{df_{ih}^m}{dt}(t) = J_{ih}[\mathbf{f}^m](t) \\ f_{ih0}^m = f_{ih}^m(t = 0)\,, \end{cases} \qquad (5.5.13)$$

for $i = 1, \ldots, n$ and $h = 1, \ldots, m$.

Solutions can be computed by means of well-known methods for integrating ordinary differential equations. Considering that a large number of sums has to be computed in J_{ih}, predictor-corrector methods have to be suggested, and Chebychev collocations seem to be convenient because of their spectral accuracy. In this case, the quadrature rule is known as Cleanshaw–Curtis rule.

In principle, one expects that the simulation should visualize, according to suitable choices of the parameters, one of the following behaviors:

a) The response of the immune system is prompt and effective, so that the tumor is completely destroyed;

b) The response of the immune system is not adequate, leading to a burst in the number of tumor cells. This can even occur after a long period of dormancy;

c) The immune response does not completely destroy the tumor. The strong reduction in the number of tumor cells induces a decay in the number of immune cells. Part of the immune cells degenerates leading to a new activation and growth of the tumor. This can either be controlled again or may lead to a final burst in the number of tumor cells.

d) The size and quality of cytokine signals, that have the possibility of activating the cells of the immune system, can effectively produce a rejection of the tumor.

Indeed, simulations show that the above qualitatively different behaviors are effectively described, and that the choice of parameters β, which refer to the interaction between tumor cells and immune system, play an important role in determining the output of the competition. In detail, it is shown that a crucial action is played by the parameter β_{21}, which represents the ability (or inability) of the tumor to induce a degenerative action of the immune cells. The system is highly sensitive to β_{21} which acts, as observed in [BEe], as a bifurcation parameter, at fixed values of $\{\beta_{12}^-, \beta_{12}^+, \gamma, \delta\}$, of the two extreme behaviors: the desired one, Case a), and the negative one, Case b).

It is interesting to notice that the competition may be long-lasting and with oscillating behavior with a final output determined by a progressive weakening of tumor cells along the competition, and activation of immune cells in Case a); in Case b), the opposite behavior is observed. The term weakening is here used to indicate that the distribution over the activation u of tumor cells gradually shifts toward low values, while activation of immune cells is used to indicate that the distribution over their activation gradually shifts toward high values. Indeed these realistic, and experimentally confirmed, behaviors described by the kinetic model are certainly encouraging toward the development of the above mentioned cooperation between mathematicians and immunologists.

The above behavior, observed in [BEe], has been the object of a systematic analysis in [STa] where the role of the parameter β_{21} is confirmed even for interaction with delay. In more details, [STa] shows how temporary weakening of the immune system may generate growth of tumors otherwise depleted by the immune system. The situation may be reversed for temporary activation of the immune system.

A further analysis is developed in [FIa] that refers to an interesting development of the model, which will be described later in the next section. In detail, the case of medically induced actions is dealt with. The action can be regarded as a control action acting for a limited time interval. Again, it is shown that the output of the competition can be modified by medically induced actions and that, for these to be effective, they have to be addressed toward modification of the parameter β_{21}.

The review of the above results may play an important role of support to experimental activity of immunologists, although it is plain that mathematics can only address the experiments and reduce their time consuming and costs. Indeed, mathematics cannot play any deeper role in the specific medical research.

5.5.3 Perspectives

Due to the extreme complexity of the physical systems we are dealing with, the model which was described in this section is, at least in some circumstances, not yet sufficient to describe the overall phenomenologic behavior of the real system. Developments are then necessary. Indeed, this topic is the object of a continuous research activity. Some developments are already available in the literature; some others simply have to be conjectured and are still under investigations. This subsection reports some conceivable developments with the aim of indicating research perspectives.

• ***Development of microscopic models***
The evolution model should be derived only after the development of appropriate models related to microscopic cell interactions. A deeper study of the cells behavior and, in particular, of cells signaling should lead to relatively more realistic models based on the true physics of the system rather than on phenomenologic interpretations. An example of how these models can be developed is reported in Chapter 6.

• ***Models with internal structure***
Models with internal structure are such that the state variable evolution equation is determined by internal as well as external actions. In fact, the evolution of the u variable cannot only be externally induced; it can also, and even, be an inner property of the cell. This important development is proposed in [Fla], in terms of model of the type

$$\frac{\partial f_i}{\partial t} + \frac{\partial}{\partial u}\left(b_i(t,u) f_i\right) = J_i[\mathbf{f}]\,, \qquad b_i = \frac{du_i}{dt}\,, \tag{5.5.14}$$

where $b_i = b_i(t,u)$ is the activation term. The simulations developed in [Fla] show that suitable external actions can effectively modify the outcome of the competition. Of course this is only a preliminary indication to immunologists. It simply shows when the competition may be able to modify the outcome of the competition in contrast with physical situations where no action is able to do it, e.g., very aggressive tumors.

• ***Models with individuals generation***
Development of models where individuals generate individuals in other populations may be related to modelling the onset of neoplastic cells to be followed by the competition with immune cells. This

framework is proposed in [ARc]. However, it still needs to be developed for real phenomena.

5.6 References

[ABa] ABBAS A.K., LICHTMANN A.H., and POBER J.S., **Cellular and Molecular Immunology**, Saunders, (1991).

[ADa] ADAM J.A. and BELLOMO N. Eds., **A Survey of Models on Tumor Immune Systems Dynamics**, Birkhäuser, Boston, (1996).

[ARa] ARLOTTI L. and BELLOMO N., On the Cauchy problem for the nonlinear Boltzmann equation, in **Lecture Notes on the Mathematical Theory of the Boltzmann Equation**, Bellomo N. Ed., World Sci., London Singapore, (1995), 1–64.

[ARb] ARLOTTI L. and LACHOWICZ M., Qualitative analysis of a nonlinear integrodifferential equation modelling tumor-host dynamics, *Mathl. Comp. Modelling - Special Issue on Modelling and Simulation Problems on Tumor–Immune System Dynamics*, Bellomo N. Ed., **23** (1996), 11–30.

[ARc] ARLOTTI L., BELLOMO N., and LACHOWICZ M., Kinetic equations modelling population dynamics, *Transp. Theory Stat. Phys.*, to appear.

[ASa] ASACHENKOV A., MARCHUK G., MOHOLER R., and ZUEV S., **Disease Dynamics**, Birkhäuser, Boston, (1994).

[BEa] BELLOMO N. and FORNI G., Dynamics of tumor interaction with the host immune system, *Mathl. Comp. Modelling*, **20** (1994), 107–122.

[BEb] BELLOMO N., FORNI G., and PREZIOSI L., On a kinetic (cellular) theory for the competition between tumors and host–immune system, *J. Biol. Systems*, **4** (1996), 479–502.

[BEc] BELLOMO N., PREZIOSI L., and FORNI G., Tumors immune system interaction: The kinetic cellular theory, in **A Survey of Models on Tumor Immune Systems Dynamics**, Adam J. and Bellomo N. Eds., Birkhäuser, Boston, (1996), 135–186.

[BEd] BELLOMO N., FIRMANI B., GUERRI G., and PREZIOSI L., On a kinetic theory of cytokine-mediated interaction between tumors and immune system, *ARI Journal, Springer*, 50 (1997), 21–32.

[BEe] BELLOMO N., FIRMANI B., and GUERRI L., Bifurcation analysis for a nonlinear system of integrodifferential equations modelling tumor immune cells competition, *Appl. Math. Letters*, **12**, (1999), 39–44.

[BEf] BELLOMO N. and DE ANGELIS, Strategies of applied mathematics toward an immuno-mathematical theory on tumors and immune system interactions, *Math. Models Meth. Appl. Sci.*, **8**, (1998), 1403–1429.

[CUa] CURTI B.D. and LONGO D.L., A brief history of immunologic thinking: It is time for Yin and Yang, in **A Survey of Models on Tumor Immune Systems Dynamics**, Adam J. and Bellomo N. Eds., Birkhäuser, Basel, (1996), 1–14.

[DEa] DEN OTTER W. and RUITENBERG E.J. Eds., **Tumor Immunology. Mechanisms, Diagnosis, Therapy**, Elsevier, New York, (1987).

[FIa] FIRMANI B., GUERRI L., and PREZIOSI L., Tumor immune system competition with medically induced activation disactivation, *Math. Models Meth. Appl. Sci.*, **9**, (1999), 491–521.

[FOa] FORNI G., FOA R., SANTONI A., and FRATI Eds., **Cytokine Induced Tumor Immunogeneticity**, Academic Press, New York, (1994).

[GRa] GREEN I., COHEN S., and MCCLUSKEY R. Eds., **Mechanisms of Tumor Immunity**, Wiley, London, New York,

(1977).

[LOa] Lo Schiavo M., Discrete kinetic cellular models of tumor immune system interactions, *Math. Models Meth. Appl. Sci.*, **6**, (1996), 1187–1210.

[MUa] Murray J., **Mathematical Biology**, Springer, Heidelberg, (1994).

[NOa] Nossal G.J., Life, death and the immune system, *Scientific American*, **269** (1993), 53–72.

[SEa] Segel L., **Modelling Dynamic Phenomena in Molecular and Cellular Biology**, Cambridge Univ. Press, Cambridge, (1984).

[STa] Stöcker S. and Curci M., Modelling and simulating the effect of cytokines on the immune response to tumor cells, *Mathl. Comp. Modelling*, **28**(1998), 1–13.

[TOa] Tomlison I. and Boomer W., Failure of programmed cell death and differentiation as causes of tumors: some simple mathematical models, *Proc. Natl. Acad. Sci. USA*, **92** (1995), 11130-11134.

[WIa] Wigner E., The unreasonable effectiveness of mathematics in natural sciences, *Comm. Pure Appl. Math.*, **13** (1960), 1–14.

Chapter 6

Kinetic Models for the Evolution of Antigens Generalized Shape

6.1 An Introduction to the Generalized Shape

This chapter shows how kinetic models of population dynamics can be developed towards the description of some interesting cellular phenomena in the immune system. Modelling at least some of the immune system features is a fascinating and challenging, although difficult, task. We feel that the generalized kinetic models that will be proposed in this chapter can contribute to this objective.

At a cellular scale, the immune system can be regarded as a system constituted by a large number of individuals, or cells, of several interacting populations. Cellular interactions modify cells behavior and, ultimately, the overall behavior of immune system itself. In fact, the evolution is determined by the statistical distribution of the interacting individuals with respect to their states.

The class of models dealt with in this chapter refers to the time evolution of the statistical distributions relative to the peculiar properties of the immune system cells. With reference to the specialized literature [SEa], [DBa]–[DBe], we recall that the fundamental variable which characterizes the binding properties of antigens, antibodies, receptors, and so on, is the so-called ***generalized shape***. Interactions at a molecular level may produce an exchange of shape that can influence the time evolution of the antigens population of each given shape. Modelling should provide the description of this time evolution.

One of the purposes of modern theoretical immunology is to give a detailed description of the physical phenomenon outlined above. A mathematical model with such an objective was proposed by Segel and Perelson [SEa]. Their approach is based on a detailed analysis of infra-molecular exchanges and interactions, and this is then transferred to modelling the rate of growth (or depletion) of the ***antigens population*** with a given ***shape***.

This chapter is organized in four sections.

Section 6.1 is this introduction.

Section 6.2 describes the mathematical model.

Section 6.3 deals with the qualitative analysis of the initial value problem.

Section 6.4 discusses some of the perspectives.

Actually, the generalized shape should be regarded as a multi-dimensional variable suitable to take into account specific features such as electric charge, position of dipole moments etc., see the discussion in [SEa]. However, the presentation of this chapter is limited to the identification of only a scalar variable. Theoretical biologists and immunologists work actively in this direction as documented in [DBd],[DBe], so that further developments of the model reported in this chapter are nowadays available and should be examined from the mathematical viewpoint. In fact, inclusion of this very specialized type of models in this book is mainly due to the fact that they can hopefully contribute to the development of immuno-mathematical theories in a framework analogous to that of the preceding chapter.

6.2 The Mathematical Model

As already mentioned in the introduction, the generalized shape is a geometrical factor, which characterizes the antigen behavior, which can be modified by interactions. The model here reported has been proposed by Segel and Perelson. Denote by u the generalized cells shape; its range is assumed to be bounded: $u \in [-L, L]$. The model is about the time evolution of the distribution function f:

$$f \,:\, (t, u) \in [0, T] \times [-L, L] \longmapsto f(t, u) \in \mathbb{R}^+ \,, \tag{6.2.1}$$

which is such that $f(t, u)\, du$ gives the expected number of antigens that, at time $t \in [0, T]$, have a shape in the range du centered at u. Hence

$$N(t) \equiv E_0[f](t) = \int_{-L}^{L} f(t, u)\, du \tag{6.2.2}$$

is the total number of antigens at time t, and

$$M[f](t) := \frac{1}{2L} \int_{-L}^{L} f(t, u)\, du \tag{6.2.3}$$

gives the average cell number per generalized shape unit. As usual, the notation $N[f]$ denotes the result of an operator N acting over f, which in some cases may be linear or even independent of f.

Following [SEa], the evolution equation is obtained by balancing the rate of change of f with the terms of growth and depletion that are modelled by carefully analyzing the causes that may determine the evolution of f for each fixed value of u. In details,

$$\frac{\partial f}{\partial t} = I[1] + S[1] + P[f] - D[f] \,, \tag{6.2.4}$$

where I is the influx from bone marrow, S the antigens source, P the reproduction of stimulated cells, D the death of unstimulated cells.

• ***Influx from bone marrow.*** The influx term I is related to the inlet, from bone marrow, of antigens with a given shape distribution. The simplest modelling is based on assuming that inlet does not depend on the antigens population density f, and that its dependence on the time and shape is given by

$$I[1](u) = \alpha\varphi(u)\,, \qquad \alpha \in \mathbb{R}_+\,.$$

• ***Source of antigens.*** The source term S can be modelled in a fashion similar to that of the inlet from bone marrow. The source can be assumed, in this case, to be time dependent:

$$S[1](t,u) = \alpha_1(t)\varphi_1(u)\,.$$

• ***Death of unstimulated cells.*** The death term D is related to the cells natural death due to lack of stimulation. Actually, only a fraction $F[f](t,u) \in [0,1]$ of cells with a certain distribution is stimulated. Hence the fraction of unstimulated cells is $(1 - F[f](t,u))$. As in [SEa], it will be assumed that D is proportional to the fraction of unstimulated cells

$$D[f](t,u) = \beta f(t,u)\,(1 - F[f](t,u))\,, \qquad (6.2.5)$$

where β is a positive constant.

In particular, Segel and Perelson proposed a ***phenomenologic model*** such that F depends on the number A of u-cells activating receptor bounds, and on the number B of u-cells disactivating receptor bounds. The model is

$$F[f](t,u) = \frac{A[f](t,u)}{\gamma + B[f](t,u) + A[f](t,u)}\,, \qquad (6.2.6)$$

where γ is a positive constant and the operators A and B are defined as follows

$$A[f](t,u) = \int_{-L}^{L} a(u,w) f(t,w)\, dw\,, \tag{6.2.7}$$

and

$$B[f](t,u) = \int_{-L}^{L} b(u,w) f(t,w)\, dw\,, \tag{6.2.8}$$

where the kernels $a, b : [-L, L]^2 \to \mathbf{R}_+$ refer to quantities which govern the activation and suppression of antigens. In particular, they are modelled as follows

$$a(u,w) = \frac{k_1}{\sqrt{2\pi}\sigma_a} \exp\left[- \frac{(u+w)^2}{2s_a^2} \right], \tag{6.2.9}$$

and

$$b(u,w) = \frac{k_2}{\sqrt{2\pi}\sigma_b} \exp\left[- \frac{(u+w)^2}{2\sigma_b^2} \right], \tag{6.2.10}$$

where k_1 and k_2 are normalization constants according to the fact that both $a(\cdot, w)$ and $b(\cdot, w)$ are regarded as probability densities for each value of w.

This assumption takes into account the fact that the fit between two shapes u and w attains its maximum when $w = -u$, which corresponds to perfectly complementary shapes.

Hence, the behavior described by this interaction model is such that, formally,

$$A \uparrow \infty \Rightarrow F \to 1\,, \qquad A \downarrow 0 \Rightarrow F \to 0\,, \tag{6.2.11}$$

meaning that the higher (lower) A is, the closer F is to one (zero).

- ***Reproduction of stimulated cells.*** The reproduction term P is modelled by a phenomenologic law which describes the fact that cellular reproduction exponentially decays as the total population increases. In addition, it can be amplified or suppressed depending on the growth of a clone, or family of clones, that may induce either self-reinforcing or suppressing effects according to a parameter λ positive and negative values respectively. The model proposed in [SEa] is the following

$$\begin{aligned} P[f](t,u) =& r_0 f(t,u) e^{\lambda f(t,u)} e^{-\eta M[f](t)} \\ & \times \frac{A[f](t,u)}{\gamma + B[f](t,u) + A[f](t,u)}, \end{aligned} \tag{6.2.12}$$

where $r_0,\ \eta > 0, \quad \lambda \in \mathbf{R},$ and $A[f]$, $B[f]$ and $M[f]$ are defined in Eqs. (6.2.3),(6.2.7)–(6.2.10).

In conclusion, the ***evolution model*** writes

$$\begin{aligned} \frac{\partial f}{\partial t}(t,u) =& \alpha\varphi(u) + \alpha_1(t)\varphi_1(u) \\ & + r_0 f(t,u) \frac{A[f](t,u)}{\gamma + B[f](t,u) + A[f](t,u)} e^{\lambda f(t,u)} e^{-\eta M[f](t)} \\ & - \beta f(t,u) \left[1 - \frac{A[f](t,u)}{\gamma + B[f](t,u) + A[f](t,u)}\right], \end{aligned} \tag{6.2.13}$$

where a time scaling, related to inlets or to encounter rates, may be included in the constants α, β, and r_0.

This class of models substantially differs from the class of models described in the preceding chapters. Common features are that the model is again an evolution model for the distribution function and that, as always in kinetic models, the microscopic picture consists in an intrinsic evolution dynamics of individuals interacting among themselves and with the environment.

However, the evolution here is not entirely defined by interactions at a microscopic level. On the contrary, significant contributions are obtained, in this model, also by heuristic phenomenologic descriptions. In addition, the evolution in this model may be modified by source and sink terms, and the nonlinearity need not be of quadratic order as those we have seen in all the models described so far.

Presently, the detailed model that rules the evolution directly follows from a phenomenologic picture of the microscopic behavior. In particular, rational and exponential functions are considered, dependent on suitable integrals weighted by the distribution function. Still, the evolution model consists of an integrodifferential equation.

6.3 On the Initial Value Problem

The initial value problem is obtained by linking the evolution equation (6.2.13) to the initial condition $f_0(\cdot)$ for the function $t \mapsto f(t, \cdot)$

$$\begin{cases} \dfrac{\partial f}{\partial t} = J[f]\,, \\ f(0, \cdot) = f_0(\cdot)\,, \end{cases} \tag{6.3.1}$$

where $J[f]$ is defined by the term on the right of Eq. (6.2.13).

The analysis of Problem (6.3.1) should examine local existence, global existence, and asymptotic behavior, of the solutions that belong to a suitable Banach space. The analysis must be related to the admissible values of those parameters that characterize the model. In particular, our attention may be focused on three classes of problems:

- Existence and uniqueness of the solutions;
- Dependence of the qualitative behavior of the solutions on the control term γ and on negative values of the parameter λ. Large positive values of λ may generate blow up of solutions.
- Linear and nonlinear stability of equilibrium points and trajectories.

Based on this analysis, computational methods may be developed in order to obtain quantitative solution of the initial value problem.

Some of the above mentioned problems are already known in the literature, and some preliminary results can be reviewed. On the other hand, several problems and, in particular, the stability analysis, are still open.

The analysis of the initial value problem was developed in [BEa] in the following function spaces:

- $\mathbf{X} = C^{\nu}([-L, L]; \mathbf{R})$: the space of real valued functions on the interval $[-L, L]$, continuous with their derivatives up to the order ν, for $\nu = 0, 1, \ldots$;
- $\mathbf{X} = L_{\infty}([-L, L]; \mathbf{R})$: the space of the essentially bounded real valued functions on the interval $[-L, L]$.

The following definitions will also be used:

- A continuous function from $[0, t_1]$ into $\mathbf{X}$ is defined to be a solution of (6.3.1) in $\mathbf{X}$ if it is continuously differentiable on $(0, t_1)$ and if its derivative equals $J[f](t, \cdot)$ for all $t \in (0, t_1)$.
- For $f \in C^{\nu}([-L, L]; \mathbf{R})$, we say that $f \geq 0$ (or that f is non-negative) if $f(u) \geq 0$ for all $u \in [-L, L]$. For $f \in L_{\infty}([-L, L]; \mathbf{R})$ we say that $f \geq 0$ (or that f is non-negative) if $f(u) \geq 0$ almost everywhere in $[-L, L]$.

Further generalizations can be found in smoother spaces, that can be convenient for computational developments, see again [BEa].

Local and global existence and uniqueness of solutions of the initial value problem (6.3.1) are stated in the following theorems, that also provide some indication on their qualitative behavior:

THEOREM 6.3.1. *Let the function* $g : \mathbf{R}^2 \to \mathbf{R}_+$ *defined by*

$$g(t, u) = I[1](u) + S[1](t, u) \tag{6.3.2}$$

be sufficiently smooth on the domain $[0, t_1] \times [-L, L]$ *for some* $t_1 > 0$*; and assume that the functions* a *and* b*, introduced in Eqs. (6.2.6) and (6.2.7), are smooth and non-negative on* $[-L, L] \times [-L, L]$*. Moreover, let* $f_0 \in X$*. Then there exists a* $t_0 \in (0, t_1]$*, which depends on the parameters of the model, such that the initial value problem (6.3.1)* *has a unique solution* $t \mapsto f(t, \cdot) \in \mathbf{X}$ *on the time interval* $[0, t_0]$. *In addition, if* $f_0 \geq 0$ *and*

$$\min_{(t,u)\in\mathcal{D}_0} g(t,u) \geq 0\,, \qquad \mathcal{D}_0 := [0, t_0] \times [-L, L] \tag{6.3.3}$$

then one has: $f(t, \cdot) \geq 0\,, \quad \forall\, t \in [0, t_0]$.

THEOREM 6.3.2. *Let the function* $g = g(t, u)$ *be sufficiently smooth on* $[0, t_1] \times [-L, L]$ *for some* $t_1 > 0$; *and assume that the functions* a *and* b *are smooth and non-negative on* $[-L, L] \times [-L, L]$. *Moreover, let*

$$f_0 \in X, \quad f_0 \geq 0, \qquad \text{and} \qquad \min_{(t,u)\in\mathcal{D}_1} g(t,u) \geq 0\,, \tag{6.3.4}$$

where $\mathcal{D}_1 := [0, t_1] \times [-L, L]$. *If* $\lambda \leq 0$ *then there exists a unique non-negative solution* $t \mapsto f(t, \cdot) \in \mathbf{X}$ *to Problem (6.3.1) on the time interval* $[0, t_1]$.

THEOREM 6.3.3. *Let the function* $g = g(t, u)$ *be sufficiently smooth on* $[0, \infty) \times [-L, L]$, *and assume that the functions* a *and* b *are smooth non-negative on* $[-L, L] \times [-L, L]$ *and such that, for some* c_0^-, $c_0^+ \in \mathbb{R}$ *and a.a.* $u, w \in [-L, L] \times [-L, L]$, *the following inequalities hold true*

$$0 < c_0^- \leq a(u, w) \leq c_0^+\,, \qquad 0 \leq b(u, w) \leq c_0^+\,. \tag{6.3.5}$$

If

$$\lambda > \eta \qquad \text{and} \qquad \min_{(t,u)\in\mathcal{D}} g(t,u) \geq 0\,, \tag{6.3.6}$$

where $\mathcal{D} := [0, \infty) \times [-L, L]$, *then the solution blows up in a finite time interval unless*

$$M[f](0) = \frac{1}{2L} \int_{-L}^{L} f(0, u)\, du$$

is sufficiently small.

The proof of Theorem 6.3.1 consists in showing that for an arbitrary choice of the model parameters, the operator J is locally Lipschitz continuous on $\mathbf{X}$, i.e., that there exists $c_\rho \in \mathbb{R}$ such that

$$\|J[f_1] - J[f_2]\| \le c_\rho \|f_1 - f_2\|, \qquad \forall\, f_1, f_2 \in B_\rho(f_0), \tag{6.3.7}$$

where $B_\rho(f_0) \subset \mathbf{X}$ is the ball with radius $\rho > 0$ and center f_0. The real number c_ρ depends on ρ and on the parameters that characterize the model. Local existence and positivity follow by immediate technical calculations.

Global existence, i.e., the proof of Theorem 6.3.2 consists in showing that for certain choices of the parameters, the solution does not grow too fast. Starting point of the proof are the inequalities

$$f(t,u) \ge f_0(u)e^{-\beta t} + \frac{1}{\beta}\left[1 - e^{-\beta t}\right] \min_{(t,u)\in\mathcal{D}_1} g(t,u), \tag{6.3.8}$$

and

$$e^{\beta t} f(t,u) \le f_0(u) + \int_0^t e^{\beta s}\left[g(s,u) + (\beta + r_0) f(s,u)\right] ds, \tag{6.3.9}$$

where β is defined in Eq. (6.2.5) and r_0 in Eq. (6.2.12). Application of Gronwall's Lemma [ZEa] yields

$$f(t,u) \le \left(\frac{\beta + r_0 + 1}{\beta + r_0}\right) f_0(u) + \left[\min_{(t,u)\in\mathcal{D}_1} g(t,u)\right] \times \left(\frac{\beta + r_0(2+\beta) + r_0^2}{\beta r_0(\beta + r_0)} + \frac{1}{\beta r_0} e^{r_0 t}\right), \tag{6.3.10}$$

for $t \in [0, t_1]$ and $u \in [-L, L]$.

Detailed analysis of the above inequalities shows that if $\lambda \leq 0$ then the right-hand side term in Eq. (6.2.13) does not grow too fast with t, and hence global existence is acquired.

On the other hand, at least for $\lambda > \eta$, the solution blows up in a finite time interval. That is, basically, Theorem 6.3.3. In fact, the proof is based on several other inequalities. In particular, if $f(t,u) \geq 0$ then

$$c_1^- M[f](t) \leq A[f](t,u) \leq c_1^+ M[f](t)\,, \tag{6.3.11}$$

and

$$c_2^- M[f](t) \leq F[f](t,u) \leq c_2^+ M[f](t)\,, \tag{6.3.12}$$

for all $u \in [-L, L]$ and where c_1^-, c_1^+, c_2^- and c_2^+ are positive constants.

Moreover, denoting by $\|\cdot\|_1$ the usual norm in the space

$$\mathbf{X} = L_1\left([-L, L]; \mathbb{R}\right)\,,$$

one has that (6.3.11) implies

$$\|J[f](t,\cdot)\|_1 \geq c_3 \left\| f(t,\cdot)e^{\lambda f(t,\cdot)} \right\|_1 \frac{M[f](t)}{c_4 + M[f](t)} e^{-\eta M[f](t)}\,, \tag{6.3.13}$$

where c_3 and c_4 are positive constants. On the other hand, convexity of the function $x \to x e^{\lambda x}$ for $x > -2/\lambda$, together with Jensen's inequality [LAa], yield

$$\left\| f(t,\cdot)e^{\lambda f(t,\cdot)} \right\|_1 \geq M[f](t) e^{\lambda M[f](t)}\,. \tag{6.3.14}$$

Hence, from Eqs. (6.3.13) and (6.3.14) it follows that

$$\|J[f](t,\cdot)\|_1 \geq c_5 \frac{(M[f](t))^2}{c_4 + M[f](t)} e^{(\lambda-\eta) M[f](t)}\,, \tag{6.3.15}$$

where c_5 is a positive constant.

Consequently if, under the conditions of the theorem, the function $f : t \in [0, t_1] \mapsto f(t, \cdot) \in \mathbf{X}$ is a non-negative solution of the initial value problem (6.3.1), then $M[f](t)$ satisfies the following inequality

$$\frac{d}{dt} M[f](t) \geq c_6 (M[f](t))^2 - \beta M[f](t) , \tag{6.3.16}$$

where c_6 is a positive constant. Let us now compare the term $M[f](t)$ with the solution of the problem

$$\frac{dy}{dt} = c_6 y^2 - \beta y , \qquad y(0) = y_0 \in \mathbb{R} , \tag{6.3.17}$$

i.e., with the function

$$y(t) = \frac{\beta}{c_6} \left[1 - \left(1 - \frac{\beta}{c_6 y_0} \right) e^{\beta t} \right]^{-1} . \tag{6.3.18}$$

It can be concluded that if

$$M[f](0) \geq \frac{\beta}{c_6} , \tag{6.3.19}$$

then the function

$$M[f](t) = \frac{1}{2L} \| f(t) \|_1 = \frac{1}{2L} N(t) \tag{6.3.20}$$

blows up in a finite time interval.

These results justify developing the necessary mathematical methods to obtain quantitative results. Again, generalized collocation methods, already outlined in Chapter 3, can be used.

6.4 Applications and Developments

The mathematical model proposed by Segel and Perelson can be regarded as a generalized kinetic equation modelling the time evolution of a statistical distribution function. The distribution is a time dependent number density function for the antigens population; it is a density function with respect to a linear geometrical factor called generalized shape.

The model has the great merit of attempting to describe a highly complicated system characterized by an inner structure that is certainly very hard to understand and classify in a way that may be desirable by modellists. The authors derived the model with a constant search for simplicity. Also due to its simplicity, it allows to describe the system behavior along simple and understandable paths. The model was followed by several further studies and technical developments mainly documented by De Boer [DEa]-[DEd].

In fact, the model can be improved. For instance, the dimensions of the variable u may be enlarged, or other models for the terms I, S, P, D may be figured. On the other hand, its generalizations seem to be of even higher interest in developing microscopic models of cellular interactions similar to those described in Chapter 5. This objective is a great challenge for the cooperation between immunologists and applied mathematicians. Once such an objective will be effectively reached, the class of models presented in Chapter 5 may be fully consistent with a detailed description of the microscopic (cellular) behavior of the system, that was there developed only at a formal level. Indeed, this appears to be the most interesting research perspective.

Referring to the qualitative analysis of the initial value problem, it has been shown above that a rather complete existence theory has been developed. Now, the problem of existence and stability of equilibrium points needs to be dealt with. A preliminary numerical analysis is shown in [SEa]. Analytic proofs that may take advantages of the analysis reported in Chapter 3 could be developed, but still should be adapted to this specific case. In particular, existence of equilibrium solutions may be analyzed in the autonomous case, that is when the source S is equal to zero. In this case, the inlet from bone marrow compensates the depletion and activation of antigens. The stationary integral equation is

$$\alpha\varphi(u) + \alpha_1\varphi_1(u) - \beta f(u)\left[1 - \frac{A[f](u)}{\gamma + B[f](u) + A[f](u)}\right]$$

$$+ r_0 f(u)\,\frac{A[f](u)}{\gamma + B[f](u) + A[f](u)}\,e^{\lambda f(u)}e^{-\eta M[f]} = 0\,. \qquad (6.4.1)$$

Linear stability is obtained by technical calculations, see [SEa]. An analysis of nonlinear stability is not available, at present, in the literature.

Referring to technical developments, one can consider the cases wherein the generalized shape is a vector or an unbounded variable. In the case $u \in (-\infty, \infty)$, the model refers to the time evolution of the distribution function

$$f\,:\,(t,u) \in [0,T] \times (-\infty,\infty) \mapsto f(t,u) \in \mathbf{R}_+\,, \qquad (6.4.2)$$

which is now such that

$$N[f](t) = \int_{-\infty}^{\infty} f(t,u)\,du\,, \qquad (6.4.3)$$

is the total number of antigens at time t. Considering that this number has to be finite, it is necessary that f shows a sufficiently rapid decay to zero when the variable u approaches infinity.

The evolution equation can be obtained by calculations similar to those reported in Section 6.2. It can be written explicitly

$$\frac{\partial f}{\partial t}(t,u) = \alpha^*\varphi(u) + \alpha_1^*(t)\varphi^*(u)$$
$$+ r_0^* f(t,u)\frac{A[f](t,u)}{\gamma^* + B[f](t,u) + A[f](t,u)}\, e^{\lambda^* f(t,u)} e^{-\eta^* N[f](t)}$$
$$- \beta^* f(t,u)\left[1 - \frac{A[f](t,u)}{\gamma^* + B[f](t,u) + A[f](t,u)}\right], \tag{6.4.4}$$

where, clearly, the operators A and B are now given by

$$A[f](t,u) = \int_{-\infty}^{\infty} a(u,w) f(t,w)\, dw\,, \tag{6.4.5}$$

and

$$B[f](t,u) = \int_{-\infty}^{\infty} b(u,w) f(t,w)\, dw\,, \tag{6.4.6}$$

and where the kernels $a = a(u,w)$ and $b = b(u,w)$ are modelled by Gaussian densities with given mean values and variances

$$a(u,w) = \frac{1}{\sqrt{2\pi}\sigma_a}\exp\left[-\frac{(u+w)^2}{2\sigma_a^2}\right], \tag{6.4.7}$$

and

$$b(u,w) = \frac{1}{\sqrt{2\pi}\sigma_b}\exp\left[-\frac{(u+w)^2}{2\sigma_b^2}\right], \tag{6.4.8}$$

respectively.

Again, this assumption takes into account that the fit of the shape w to the shape u decreases to zero at infinity, and attains its maximum value when $w = -u$, which corresponds to perfectly complementary shapes.

The evolution model has the same structure of Eq. (6.2.13), where $M[f]$ is replaced by $N[f]$ and where the constants of the model, which are here denoted by $\alpha^*, \beta^*, \gamma^*, r_0^*, \eta^*$ and λ^*, may differ from those of the model reported in Section 6.2.

Finally we remark that, assuming that a modelling of the individual cellular behavior can be given, then the model can be generalized to several interacting populations and may be a candidate to model various aspects of the immune system behavior. A development, that appears to be of particular interest, consists in describing this model as a sub-model of a larger one. For instance, the model that has been treated in Chapter 5 may take advantage of the description of cellular interactions provided by the present one. This aspect is analogous to using the coagulation-fragmentation models dealt with in Chapter 4 as support to other models.

6.5 References

[BEa] BELLOMO N., LACHOWICZ M., and POLEWCZAK J., On the evolution of the generalized shape in immunology: Qualitative analysis of the solutions to the Segel-Perelson model, *Appl. Math. Letters*, **10** 53–58, (1997).

[DBa] DE BOER R. J. and HOGEWEG P., Tumor escape from immune elimination: Simplified precursor bound cytotoxicity models, *J. Theoretical Biol.*, **113**, (1985), 331–351.

[DBb] DE BOER R. J. and HOGEWEG P., Immunological discrimination between self and non-self by precursor depletion and memory accumulation, *J. Theoretical Biol.*, **124**, (1987), 343–369.

[DBc] DE BOER R. J. and PERELSON S., Size and connectivity as emergent properties of a developing immune network, *J. Theoretical Biol.*, **149** (1991), 381–429.

[DBd] DE BOER R. J., SEGEL L., and PERELSON S., Pattern formation in one or two dimensional shape-space models of the immune system, *J. Theoretical Biol.*, **155** (1992), 295–333.

[DBe] DE BOER R. J., BOERLIJST M. C., SULZER B., and PERELSON S., A new bell shaped functions for idiotypic interactions based on cross linking, *Bull. Math. Biol.*, **58** (1996), 285–312.

[LAa] LAKSHMIKANTHAN V. and LEELA S., **Differential and Integral Inequalities**, Academic Press, New York, (1969).

[SEa] SEGEL L. A. and PERELSON A. S., Computations in shape space: A new approach to immune network theory, in **Theoretical Immunology II, SFI Studies in the Science of Complexity**, Perelson A. S. Ed., Addison-Wesley, Reading (Mass), (1988), 321–342.

[ZEa] ZEIDER E., **Applied Functional Analysis**, Springer, Heidelberg, 1995.

Chapter 7

The Boltzmann Model

7.1 Introduction

A fluid is a disordered system of interacting particles moving in all directions and contained in a space domain $\Lambda \subset \mathbb{R}^3$, possibly equal to the whole space $\mathbb{R}^3$. When the position of each particle is correctly identified by the coordinates, say $\mathbf{x}_k$ for the k-th particle, $k = 1, \ldots, N$, of its center of mass, the system may be reduced to a set of point masses and conveniently referred to a fixed frame of orthogonal axes. This is the case, for instance, when the particles have spherically symmetric shapes, and the rotational degrees of freedom can be ignored.

When the domain Λ is bounded, the particles interact with its walls $\partial\Lambda_w$. In addition, if Λ contains obstacles, that is to say other space domains Λ^* contained in Λ, then the particles also interact with the walls of $\partial\Lambda^*$ of Λ^*. In some instances, e.g., for flows around space vehicles, the obstacles are objects scattered throughout the space $\mathbb{R}^3$.

It is generally believed in physics, and in statistical mechanics in particular, that a complete understanding of the macroscopic properties of a fluid certainly follows from the detailed knowledge of the state of each one of its atoms or molecules. In most fluids of practical interest, these states and their evolution are ruled by the laws of classical mechanics which, for a system of N particles, consist in the following set of ordinary differential equations

$$\begin{cases} \dfrac{d\mathbf{x}_k}{dt} = \mathbf{v}_k \\[2ex] \dfrac{d\mathbf{v}_k}{dt} = \mathbf{F}_k = \mathbf{f}_k + \displaystyle\sum_{k'=1}^{N} \mathbf{f}_{k'k} , \end{cases} \tag{7.1.1}$$

$k = 1, \ldots, N$, with initial conditions

$$\mathbf{x}_k(0) = \mathbf{x}_{k0} , \quad \mathbf{v}_k(0) = \mathbf{v}_{k0} , \qquad k = 1, \ldots, N , \tag{7.1.2}$$

where $\mathbf{F}_k$ is the force, referred to the mass m_k, acting on the k-th particle. $\mathbf{F}_k$ can be expressed by the superposition of an external field $\mathbf{f}_k$ and of forces such as $\mathbf{f}_{k'k}$ acting on the k-th particle due to the presence of the k'-th one among the other particles of the system.

This approach prescribes that Eqs.(7.1.1) can be integrated, at least in principles, and that the macroscopic properties of the fluid can be obtained as ensemble averages of the microscopic variables. It is very hard, however, to implement this program unless suitable simplifications are introduced. Indeed, the large values of the total number N of molecules (about 10^{20} for a gas in normal conditions), and unavoidable inaccuracies in the knowledge of initial conditions result in the impossibility of retrieving and manipulating the microscopic information contained in the total set of state vectors $\{\mathbf{x}_k, \mathbf{v}_k\}_{k=1}^{N}$. Therefore problems arise in taking advantage of Eqs. (7.1.1) and (7.1.2) to compute the time and space evolution of macroscopic observables such as the mass density

$$\rho = \rho(t, \mathbf{x}) ,$$

mass velocity

$$\mathbf{u} = \mathbf{u}(t, \mathbf{x}) ,$$

energy

$$\mathcal{E} = \mathcal{E}(t, \mathbf{x}),$$

and stress tensor

$$\mathbf{P} = \mathbf{P}(t, \mathbf{x}) = [p_{ij}(t, \mathbf{x})], \quad i, j = 1, 2, 3.$$

A possible approach is the classical one of fluid dynamics [CMa], which consists in deriving the evolution equations, related to the above observables, under several *a priori* assumptions such as the hypothesis of continuity of matter (continuum assumption). This constitutes a good approximation of a real system if the mean distance between the pairs of particles is small when compared with the characteristic lengths of the system, e.g., the typical length of Λ or of Λ^*. Yet, if intermolecular distances are of the same order of magnitude of these lengths, the continuum hypothesis does not hold, and a discrepancy is expected between the continuum fluid dynamics description and that one furnished by Eqs. (7.1.1).

On the other hand, direct use of Eq. (7.1.1) to recover the macroscopic observables is an almost impossible task. This is not only related, as mentioned above, to the difficulty of dealing with such a large system of ordinary differential equations, but also to the subsequent difficulty of computing the limits which lead to the macroscopic quantities. For instance, the expected mass density $E_0[\rho]$ for a system of identical particles of mass m, can be approximated by the ratio

$$E_0[\rho] = m \, \frac{\Delta N}{\Delta \mathbf{x}} \, . \tag{7.1.3}$$

Macroscopic limits should be found when the volume $\Delta \mathbf{x}$ tends to zero although containing a large number of particles. Fluctuations cannot be avoided.

Additional difficulties are related to the computation of the other macroscopic variables. Therefore constitutive relations are needed. This argument is extensively discussed in the introductory chapters of

[CEb], where an historical background of this branch of mathematical physics is also given.

Hence, considering that it is not possible to deal either with the equations of continuum fluid dynamics [CMa] or with those of single particles dynamics (7.1.1), a different model is needed. Boltzmann's idea was to introduce the one particle statistical distribution function

$$f : (t, \mathbf{x}, \mathbf{v}) \in [0, T] \times \Lambda \times \mathbb{R}^3 \mapsto f(t, \mathbf{x}, \mathbf{v}) \in \mathbb{R}_+ , \tag{7.1.4}$$

where Λ is a connected subset of $\mathbb{R}^3$, and to derive an evolution equation for such a distribution. Indeed if f is known and such that

$$v^n f \in L_1(\Lambda \times \mathbb{R}^3) , \qquad n = 0, 1, 2 , \quad t \in [0, T] , \tag{7.1.5}$$

then the main macroscopic observables can be computed. In particular, mass density is given by

$$\rho(t, \mathbf{x}) \;\equiv\; m\ N(t, \mathbf{x}) = m \int_{\mathbb{R}^3} f(t, \mathbf{x}, \mathbf{v})\, d\mathbf{v} , \tag{7.1.6}$$

mass velocity is given by

$$\mathbf{u}(t, \mathbf{x}) = \frac{m}{\rho(t, \mathbf{x})} \int_{\mathbb{R}^3} \mathbf{v}\, f(t, \mathbf{x}, \mathbf{v})\, d\mathbf{v} , \tag{7.1.7}$$

and the mean translational energy of a monoatomic perfect gas is

$$\mathcal{E}(t, \mathbf{x}) = \frac{1}{3(k/m)\rho(t, \mathbf{x})} \int_{\mathbb{R}^3} \left[\mathbf{v} - \mathbf{u}(t, \mathbf{x})\right]^2 f(t, \mathbf{x}, \mathbf{v})\, d\mathbf{v} , \tag{7.1.8}$$

where k is the Boltzmann constant. The mean energy can be related to the temperature by suitable assumptions on the thermodynamics of the system.

It is clear that this type of modelling may only describe a rough approximation of physical reality. Indeed, the state of an N-particle gas is statistically described by the N-particle distribution function

$$f_N = f_N(t, \mathbf{x}_1, \mathbf{v}_1, \ldots \mathbf{x}_k, \mathbf{v}_k, \ldots, \mathbf{x}_N, \mathbf{v}_N) \,. \tag{7.1.9}$$

The one particle distribution function (see Example 2.3.13) may be obtained by the marginal density of the N-particle distribution function. A rigorous derivation of the evolution equation leads to a hierarchy of equations, the BBGKY hierarchy, involving all distribution from the first to the last. Then the evolution equation for $f = f_1$ is only an approximation, however useful, of physical reality.

All the above comments show that the Boltzmann equation is a model that only approximates physical reality. In other words, it is based on several approximations, that cannot be rigorously justified.

The application of Boltzmann equation to fluid dynamics starts with the formulation of mathematical problems, generally an initial and/or boundary value problem. Then it goes through developing a qualitative theory related to existence, regularity and asymptotic behavior of the solutions. Finally it ends up with the application of suitable computational techniques to evaluate quantitative solutions. The above project presents several difficulties hard to be tackled, and that should be regarded as only partially solved.

Considering that we are interested in the generalization of Boltzmann equation to fields generally different from that of molecular gas dynamics, this chapter provides only a review of the main mathematical results on this particular equation. They concern stating the problem, existence theory, and computational methods, all oriented towards fluid dynamical applications. The aim is to provide a concise guide to the literature that can be related to, and hopefully developed towards, the qualitative analysis of generalized Boltzmann models.

In addition, our discussion here is limited to the statement of the mathematical problems and to a survey of the main technical results. A concise sketch of the proof is only occasionally given. Interested readers are referred to Chapter 1 of [BLf], or the book by Glassey [GLa] for the analysis of the Cauchy problem; to the book by Maslowa [MSc] and to paper [AKg] for boundary value problems; to Chapter 2 of [BLf] for the asymptotic theory; to the survey papers [BLg] and [GNa], [GNb] for computational schemes. A complete bibliography and further details on the proofs can be recovered in the above cited references.

In more details, this chapter is divided into five sections.

Section 7.1 is this introduction.

Section 7.2 contains a concise description of the Boltzmann equation as a mathematical model of the phenomenologic non-equilibrium kinetic theory of fluids.

Section 7.3 deals with the mathematical formulation of problems: initial and initial-boundary value problems, statement of shock waves, kinetic equations towards asymptotic theory.

Section 7.4 provides a survey of analytic results concerning the problems stated in Section 7.2.

Section 7.5 consists in a survey of the literature about computational problems.

The analysis of the Cauchy problem for large values of the time parameter has been, and still is, a hard and challenging topic for applied mathematicians. Despite the various efforts spent to obtain a well developed existence theory, mainly based upon the fundamental papers by DiPerna and Lions [DPa], [DPb] and Lions [LNa]-[LNc], the whole matter cannot be regarded as satisfactorily treated. As we shall see, uniqueness can be obtained only for special and sufficiently small initial conditions, while existence theorems are often not directly useful for computational analysis.

Even greater difficulties have to be faced in the computational treatment of fluid dynamical problems. This is also related to the fact that only a limited number of analytic results can be exploited for the applications. Considering that this book deals with the generalizations of the Boltzmann equation, we will not report about all the results of the cited surveys. Our aim is only to point out the difficulties related to nonlinear kinetic models, and make reference to the link between existence theorems and computational treatment of evolution problems. Hopefully, the reader interested in generalizing the Boltzmann equation will take some advantage of this background.

7.2 The Nonlinear Boltzmann Equation

The Boltzmann equation is a mathematical model for the evolution of the statistical distribution function of a diluted gas. The original derivation, due to Boltzmann, can be regarded as a heuristic (phenomenological) derivation based upon several assumptions that cannot be justified in a rigorous framework. In fact, one has to look at these hypotheses as mere approximations of physical reality, and hence the model can only roughly describe the system behavior. All the same, as it is well known, several attempts have been developed in order to rigorously justify the Boltzmann model. This matter is well documented in Chapter 4 of [CEb] and in the bibliography therein cited. In this section we will report about the main features of phenomenological derivation. Once more, the presentation will be concise while bibliographical indications are given for those readers interested in enlarging their knowledge on this subject. The lines followed to derive Boltzmann equation, on the other hand, are quite similar to those applied in the derivation of generalized kinetic models.

Recall that the distribution function f, assumed to be a smooth non-negative function on $\mathbb{R}^+ \times \mathbb{R}^3 \times \mathbb{R}^3$, is such that

$$f(t, \mathbf{x}, \mathbf{v}, t)\, d\mathbf{x}\, d\mathbf{v} \tag{7.2.1}$$

gives the expected number of particles in the elementary volume $d\mathbf{x}\, d\mathbf{v}$ centered at the phase point $(\mathbf{x}, \mathbf{v})$. The time dependence in Boltzmann model allows that, as time goes, the number of particles inside $d\mathbf{x}\, d\mathbf{v}$ may change if the system is away from equilibrium. Obviously, such a change is computed by balancing the incoming and outgoing particles in $d\mathbf{x}\, d\mathbf{v}$.

The idea behind this balance of losses and gains in $d\mathbf{x}\, d\mathbf{v}$ is that particles are lost or gained in $d\mathbf{x}$ by free streaming, while they leave or enter $d\mathbf{v}$ as a result of collisions with other particles. The size of the volume element $d\mathbf{x}\, d\mathbf{v}$ must be on one hand so large that the number of particles contained in it justifies the use of statistical methods, and on the other hand so small that information contained in it have local character.

Clearly, these two features are not compatible in general, hence problems are expected in justifying the whole procedure. Hopefully, in the cases of practical interest, the molecule size does fall in a range of values which are small when compared to those of the volume elements $d\mathbf{x}$ which, in turn, can be considered as microscopic with respect to the observations scale. Mathematically, this is achieved in the Boltzmann-Grad limit, which allows the number of particles N tends to infinity, and the radius of action σ tends to zero, in such a way that

$$\sigma N \to 0 \qquad \text{and} \qquad \sigma N^2 \to c \in (0, \infty)\,. \tag{7.2.2}$$

In this approach, it is also assumed that in an infinitesimally short time interval, during which molecules with speed $\mathbf{v}$ cannot cross the whole of $d\mathbf{x}$, a large number of collisions takes place inside $d\mathbf{x}\, d\mathbf{v}$.

Some further assumptions are needed:

Assumption 7.2.1. *The total number of collisions between particles with velocity* $\mathbf{v}$ *and particles with velocity* $\mathbf{w}$*, in the time* dt*, equals the product between the elementary volume* $\mathcal{B}(\mathbf{n}, \mathbf{q})d\mathbf{n}$ *of relevant collision cylinders, times the number of particles with velocity* $\mathbf{w}$ *at point* $\mathbf{x}$*:*

$$\mathcal{B}(\mathbf{n}, \mathbf{q})\, d\mathbf{n}\, f(t, \mathbf{x}, \mathbf{w})\, d\mathbf{w}\,, \tag{7.2.3}$$

where $\mathbf{q} = \mathbf{w} - \mathbf{v}$, *and where* $\mathbf{n}$ *is the unit vector outgoing from the interaction sphere.*

Assumption 7.2.2. *Collisions take place between uncorrelated particles, that is (chaos assumption): for all* $\mathbf{x}, \mathbf{y} \in \Lambda$ *and* $\mathbf{v}, \mathbf{w} \in \mathbb{R}^3$

$$f_2(t, \mathbf{x}, \mathbf{y}, \mathbf{v}, \mathbf{w}) d\mathbf{x}\, d\mathbf{y}\, d\mathbf{v}\, d\mathbf{w} = f(t, \mathbf{x}, \mathbf{v}) d\mathbf{x}\, d\mathbf{v}\ f(t, \mathbf{y}, \mathbf{w}) d\mathbf{y}\, d\mathbf{w}\,. \tag{7.2.4}$$

Assumption 7.2.3. *Collisions involving more than two particles can be neglected.*

Assumption 7.2.4. *Variations inside* $d\mathbf{x}$ *of the distribution function are irrelevant. Therefore,* f *evaluated at* $\mathbf{x}$ *may be replaced by* f *evaluated at the position of the field particle.*

Once such an approach has been accepted, one finds that f must obey the balance equation

$$\begin{aligned} &\text{Time derivative of} \quad f\, d\mathbf{x}\, d\mathbf{v} = (\text{flow in } d\mathbf{x}) - (\text{flow out } d\mathbf{x}) \\ &+ (\text{collisions } \mathbf{v}' \neq \mathbf{v} \to \mathbf{v}) - (\text{collisions } \mathbf{v} \to \mathbf{v}' \neq \mathbf{v}) \end{aligned} \tag{7.2.5}$$

The net flow of particles through the volume element $d\mathbf{x}$ in a time dt is easily shown to be

$$-\, d\mathbf{v}\, dt\, d\mathbf{x}\, \langle \mathbf{v}\,,\, \nabla_{\mathbf{x}} \rangle\, f(t, \mathbf{x}, \mathbf{v})\,. \tag{7.2.6}$$

The remaining two terms contain the number of collisions that test particles undergo in the time dt. We consider a collision to be the only event which may modify the velocities of the colliding particles. Moreover, the time interval dt is assumed to be so short with respect to the average collision time that events consisting of two or more collisions in dt may be neglected.

Combining what have been stated yields the expression of the celebrated Boltzmann model

$$\left(\frac{\partial}{\partial t} + \langle \mathbf{v}\,,\, \nabla_{\mathbf{x}} \rangle + \langle \mathbf{F}\,,\, \nabla_{\mathbf{v}} \rangle \right) f = J[f] = J_1[f] - J_2[f]\,, \tag{7.2.7}$$

where

$$J_1[f](t,\mathbf{x},\mathbf{v}) = \int_{\Lambda\times\mathbf{S}_+^2} \mathcal{B}(\mathbf{n},\mathbf{q}) f(t,\mathbf{x},\mathbf{v}') f(t,\mathbf{x},\mathbf{w}')\, d\mathbf{n}\, d\mathbf{w}, \qquad (7.2.8)$$

$$J_2[f](t,\mathbf{x},\mathbf{v}) = f(t,\mathbf{x},\mathbf{v}) \int_{\Lambda\times\mathbf{S}_+^2} \mathcal{B}(\mathbf{n},\mathbf{q}) f(t,\mathbf{x},\mathbf{w})\, d\mathbf{n}\, d\mathbf{w}, \qquad (7.2.9)$$

and where $\mathbf{v}'$ and $\mathbf{w}'$ are explicit functions of $\mathbf{v}$ and $\mathbf{w}$ deduced by the detailed collision mechanics. Moreover: $\mathbf{F}$ is the external force field acting on the particles; $\mathcal{B}$ is a collision kernel which depends upon the interaction potential; and $\mathbf{S}_+^2$ is the surface of unit radius of the vectors $\mathbf{n}$ that point away from the collision.

At this stage, the determinism of the particle dynamics is lost, and it has been replaced by an average over all possible initial positions and velocities.

With reference to the specialized literature, we are now interested in reporting some fundamental properties of the Boltzmann model. In particular, we will concentrate our attention upon existence and stability of equilibrium solutions. Recall that

$$J[f] = 0 \qquad (7.2.10)$$

is a functional equation which admits the so-called Maxwellian equilibrium solution $\omega(t,\mathbf{x},\mathbf{v})$ given by

$$\omega(t,\mathbf{x},\mathbf{v}) = a(t,\mathbf{x}) \exp\left(-\frac{[\mathbf{v}-\mathbf{b}(t,\mathbf{x})]^2}{c(t,\mathbf{x})}\right), \qquad (7.2.11)$$

where the terms $a(t,\mathbf{x})$, $\mathbf{b}(t,\mathbf{x})$ and $c(t,\mathbf{x})$ are related to the macroscopic observables.

Trend towards equilibrium is described by the entropy functional

$$H[f](t) = \int f(t, \mathbf{x}, \mathbf{v}) \log f(t, \mathbf{x}, \mathbf{v}) \, d\mathbf{x} \, d\mathbf{v} \tag{7.2.12}$$

under the assumption that the term $f \log f$ is integrable. In fact it may be shown that

$$H[f](t_1) \geq H[f](t_2) \,, \quad \forall\, t_1, t_2 \,, \quad 0 \leq t_1 < t_2 \,, \tag{7.2.13}$$

and that equality holds at equilibrium.

7.3 Mathematical Problems

Consider at first the ***initial value problem*** for the Boltzmann equation in the whole space $\mathbb{R}^3$ and in the ***absence of an external force field***, and hence for $\mathbf{v} = const.$ The problem is stated when the evolution equation (7.2.7) is linked to a given initial condition

$$f_0(\mathbf{x}, \mathbf{v}) = f(0, \mathbf{x}, \mathbf{v}) \geq 0 \,, \qquad (\mathbf{x}, \mathbf{v}) \in \mathbb{R}^6 \,, \tag{7.3.1}$$

which will be assumed to decay to zero as $|\mathbf{x}| \to \infty$.

The integral form of the initial value problem for the Boltzmann equation may be found by direct integration over time, and it reads

$$f^{\#}(t, \mathbf{x}, \mathbf{v}) = f_0(\mathbf{x}, \mathbf{v}) + \int_0^t J^{\#}(s, \mathbf{x}, \mathbf{v}) \, ds \,, \tag{7.3.2}$$

where

$$f^{\#}(t, \mathbf{x}, \mathbf{v}) := f(t, \mathbf{x} + \mathbf{v}\, t, \mathbf{v}) \,, \tag{7.3.3}$$

and

$$J^{\#}(t, \mathbf{x}, \mathbf{v}) := J[f](t, \mathbf{x} + \mathbf{v}\, t, \mathbf{v}) \,. \tag{7.3.4}$$

A ***solution*** to the ***initial value problem*** may be correctly defined only if a suitable Banach space, say $\mathbf{X}$, is specified. Then, the evolution problem (7.3.2) may be written in abstract form as

$$f = U[f]\,, \tag{7.3.5}$$

where $U : \mathbf{X} \to \mathbf{X}$ is the operator defined by

$$U[f](t,\mathbf{x},\mathbf{v}) := f_0(\mathbf{x}-\mathbf{v}t,\mathbf{v}) + \int_0^t J[f](s,\mathbf{x}-\mathbf{v}(t-s),\mathbf{v})\,ds\,. \tag{7.3.6}$$

However, even without specifying, for the moment, the details of the function space $\mathbf{X}$, the following definitions can still be given:

Definition 7.3.1. *A function $t \mapsto f(t,\mathbf{x},\mathbf{v})$ is a **mild solution** of the initial value problem for the Boltzmann equation if $f(t,\cdot,\cdot)$ belongs to $\mathbf{X}$ for all $t \in [0,T]$ and Eq.(7.3.2), or Eq.(7.3.5), is satisfied in the classical sense, i.e., pointwise.*

Definition 7.3.2. *A function $t \mapsto f(t,\mathbf{x},\mathbf{v})$ is a **classical solution** of the initial value problem for the Boltzmann equation if $f(t,\cdot,\cdot)$ belongs to $\mathbf{X}$ for all $t \in [0,T]$, if $(t,\mathbf{x}) \mapsto f(t,\mathbf{x},\cdot)$ is continuously differentiable, and if $t \mapsto f(t,\cdot,\cdot)$ satisfies Eq.(7.3.2) in the classical sense.*

Definition 7.3.3. *A function $t \mapsto f(t,\mathbf{x},\mathbf{v})$ is a **renormalized solution** of Eq. (7.3.2) if*

$$\frac{1}{1+f}\,J_i[f] \in L_{1,loc}\left([0,\infty)\times \mathbb{R}^3 \times \mathbb{R}^3\right)\,, \qquad i = 1,2\,, \tag{7.3.7}$$

and if $g = \log(1+f)$ is a solution of the equation

$$\frac{\partial g}{\partial t} + \langle \mathbf{v}\,,\,\nabla_{\mathbf{x}}\rangle g = \frac{1}{1+f}J[f]\,, \tag{7.3.8}$$

in the sense of distributions.

To state the ***initial boundary value problems***, and the ***boundary value problems***, gas–surface interaction phenomena need to be modelled. In particular, two classes of problems can be considered among several others:

- ***Interior domain problems***: those related to a gas which is contained in a volume with a fixed solid boundary;
- ***Exterior domain problems***: arising when the gas occupies the whole space $\mathbb{R}^3$ which, however, contains some obstacles.

In both cases the surface of the solid wall is defined by $\partial\Lambda$, and the normal to the surface, directed towards the gas, is denoted by $\boldsymbol{\nu}$.

To define the boundary conditions on a solid wall, we need to introduce the partial incoming and outgoing traces f^+ and f^- on the boundary $\partial\Lambda$ which, for continuous f, can be defined as follows

$$\begin{cases} f^+(\mathbf{x},\mathbf{v}) = f(\mathbf{x},\mathbf{v})\,, & \mathbf{x}\in\partial\Lambda\,, \quad \langle \mathbf{v}\,,\,\boldsymbol{\nu}(\mathbf{x})\rangle > 0 \\ f^+(\mathbf{x},\mathbf{v}) = 0\,, & \mathbf{x}\in\partial\Lambda\,, \quad \langle \mathbf{v}\,,\,\boldsymbol{\nu}(\mathbf{x})\rangle \le 0\,, \end{cases} \tag{7.3.9}$$

and

$$\begin{cases} f^-(\mathbf{x},\mathbf{v}) = f(\mathbf{x},\mathbf{v})\,, & \mathbf{x}\in\partial\Lambda\,, \quad \langle \mathbf{v}\,,\,\boldsymbol{\nu}(\mathbf{x})\rangle < 0 \\ f^-(\mathbf{x},\mathbf{v}) = 0\,, & \mathbf{x}\in\partial\Lambda\,, \quad \langle \mathbf{v}\,,\,\boldsymbol{\nu}(\mathbf{x})\rangle \ge 0\,. \end{cases} \tag{7.3.10}$$

Then, the boundary condition can be formally defined as follows

$$f^+(t,\mathbf{x},\mathbf{v}) = Q[f^-](t,\mathbf{x},\mathbf{v})\,, \tag{7.3.11}$$

where the operator Q, from $L_{1,loc}\left([0,\infty)\times\mathbb{R}^3\times\mathbb{R}^3\right)$ into itself, maps the distribution function of particles colliding with the surface into the distribution function of particles leaving it. For a broad range of physical problems, the operator Q is characterized by the following properties:

1. *Q is linear, of local type with respect to* $\mathbf{x}$, *and positive*

$$f \geq 0 \quad \Rightarrow \quad Q[f] \geq 0 \,. \tag{7.3.12}$$

2. *Q preserves the mass; i.e., the flux of incoming particles equals the flux of particles leaving the surface.*

3. *Q preserves local equilibrium at the boundary*

$$\omega_w^+ = Q[\omega_w^-] \,, \tag{7.3.13}$$

where ω_w is the Maxwellian temperature distribution on the wall.

4. *Q is dissipative; i.e., Q satisfies the Darrozes and Guiraud inequality on the wall*

$$\int_{\mathbb{R}^3} \langle \mathbf{v} \,,\, \nu \rangle \left(f^- + Q[f^-] \right) \left(\log f + \frac{|\mathbf{v}|^2}{\mathcal{E}} \right) d\mathbf{v} \leq 0 \,. \tag{7.3.14}$$

The book by Cercignani [CEa] provides a description of some specific models. Nonlinear boundary conditions are studied in [HAb].

In the case of ***interior domains***, the formulation of initial-boundary value problems consists in linking the evolution equation (7.3.2) and its initial conditions to the boundary condition (7.3.11) on the wall $\partial\Lambda$. In the case of ***exterior domains***, suitable Maxwellian equilibrium conditions at infinity are generally added to the boundary conditions on the wall of the obstacles. In all cases, boundary conditions must be linked to the steady Boltzmann equation if boundary value problems are considered, their solutions being defined analogously to those of the initial value problem.

A typical boundary value problem is the analysis of ***shock wave profiles*** in one space dimension. It consists in finding a smooth profile $f : (x, \mathbf{v}) \in \mathbb{R} \times \mathbb{R}^3 \mapsto f(x, \mathbf{v}) \in \mathbb{R}_+$ which satisfies the steady

Boltzmann equation and the asymptotic conditions

$$\lim_{x \to -\infty} = \omega^-(\mathbf{v}) \,, \qquad \lim_{x \to +\infty} = \omega^+(\mathbf{v}) \,, \tag{7.3.15}$$

where ω^- and ω^+ are two prescribed Maxwellians.

In addition, the shock wave profile is required to be the limit, as t tends to infinity, of the initial value problem solution with initial conditions

$$\begin{cases} x \leq 0 : & f_0(x, \mathbf{v}) = \omega^-(\mathbf{v}) \\ x > 0 : & f_0(x, \mathbf{v}) = \omega^+(\mathbf{v}) \quad . \end{cases} \tag{7.3.16}$$

All mathematical formulations we have just seen refer to the Boltzmann equation in the absence of an external field. When an external field acts on the particles, the trajectories are identified by the solutions of ***the free particle equation of motion***:

$$\begin{cases} \dfrac{d\mathbf{x}}{dt} = \mathbf{v} \,, \quad \mathbf{x}_0 = \mathbf{x}(t = 0) \\ \dfrac{d\mathbf{v}}{dt} = \mathbf{F}(t, \mathbf{x}, \mathbf{v}) \,, \quad \mathbf{v}_0 = \mathbf{v}(t = 0) \,. \end{cases} \tag{7.3.17}$$

Such initial value problem is generally based on the assumption that the force field assures existence and uniqueness of its solutions

$$\mathbf{x} = \mathbf{x}(t, \mathbf{x}_0, \mathbf{v}_0) \,, \qquad \mathbf{v} = \mathbf{v}(t, \mathbf{x}_0, \mathbf{v}_0) \,.$$

Hence, the integral formulation of the initial value problem (7.2.7) with (7.3.1) becomes

$$f(t, \mathbf{x}, \mathbf{v}) = f_0\left(\mathbf{x}(-t, \mathbf{x}, \mathbf{v}), \mathbf{v}(-t, \mathbf{x}, \mathbf{v})\right) + \int_0^t J[f]\left(s, \mathbf{x}(s - t, \mathbf{x}, \mathbf{v}), \mathbf{v}(s - t, \mathbf{x}, \mathbf{v})\right) ds \,. \tag{7.3.18}$$

When dealing with computational problems in fluid dynamics, it may be useful, or even necessary, to partition the volume occupied by the gas into two families of subsets: the domains where one can apply continuum fluids equations, and the domains where the Boltzmann equation is of use. This decomposition has several advantages, but involves additional difficulties concerning the necessary matching between the two models. In particular, the hydrodynamic limit is defined and, correspondingly, kinetic equations obtained, that are referred to as ***asymptotic theories of the Boltzmann equation***. These may correctly be developed only if the Boltzmann equation is written in suitable dimensionless form, therefore the following numbers are introduced

- the Knudsen number

$$K_n = \frac{\ell}{d}\,,$$

  where ℓ is the mean free path and d is a characteristic macroscopic scale;

- the Mach number

$$M_a = \frac{|\mathbf{u}|}{c_s}\,,$$

  where c_s is the sound speed and $|\mathbf{u}|$ the macroscopic mass velocity;

- the Reynolds number

$$R_e = \frac{\rho|\mathbf{u}|d}{\mu}\,,$$

  where μ is the viscosity coefficient;

- the Strouhal number

$$S_h = \frac{d}{|\mathbf{u}|\tau}\,,$$

  where τ refers to the macroscopic time scale.

So, the Knudsen number is proportional to the ratio between the Mach and the Reynolds numbers.

Different scalings provide, as documented in [LAc], different non-dimensional models, all exhibiting a singular dependence on a parameter. Two typical cases can be considered. One is

$$\frac{\partial f}{\partial t} + \langle \mathbf{v}\,, \nabla_{\mathbf{x}} \rangle f = \frac{1}{\varepsilon}\, J[f]\,, \tag{7.3.19}$$

which corresponds to $M_a = S_h = 1$ and where ε is proportional to the Knudsen number. Or, alternatively

$$M_a \frac{\partial f}{\partial t} + \langle \mathbf{v}\,, \nabla_{\mathbf{x}} \rangle f = \frac{1}{\varepsilon}\, J[f]\,, \tag{7.3.20}$$

where both M_a and ε are small parameters. The analysis is developed under the assumption that both M_a and ε are of the same order of magnitude, and both much lower than one.

The singular perturbation analysis aims to recover the hydrodynamic equations, and to overlap the kinetic equations with those of continuum hydrodynamics, as the Knudsen number tends to zero.

7.4 Analytic Treatment

This section provides a survey of the mathematical results on the qualitative analysis, existence theory and asymptotic behavior of the solutions to the initial and initial-boundary value problems for the spatially inhomogeneous Boltzmann equation. Additional problems that will be reviewed are shock wave problems and asymptotic analyses towards the hydrodynamic limit. We shall first refer to the equation in the ***absence of an external force field***; afterwards, some generalizations will be discussed when a force field is present. Moreover, since the literature on the spatially homogeneous problem is already reported in [TRa], and refers to the work essentially developed following the papers by Arkeryd [AKa], [AKb], here we shall focus our analysis only on space depending problems. Finally, we shall not insist upon local existence, that can be proven in several ways by classical methods of functional analysis: fixed point theorems, semigroup theory, or the so-called Kaniel and Shinbrot iterative scheme [KAa]. A complete information on this topic may be found on the review paper by Greenberg, Polewczak and Zweifel [GRa].

7.4.1 The Cauchy problem for small initial data

The analysis of the nonlocal solution of the initial value problem provided, in the last years, several interesting results concerning existence of solutions, uniqueness and, at least in some cases, asymptotic behavior. The mathematical literature can be organized in three main fields:

- Solutions for small, in the L_1-norm, initial data. This type of results refers to existence and uniqueness of mild solutions for initial conditions close either to vacuum, or to local Maxwellians, or to the solution of the spatially homogeneous case. In the case of initial conditions close to a global Maxwellian (i.e., a distribution taken with constant parameters), one studies existence of solutions together with their asymptotic trend to equilibrium.
- Solutions for large, in the L_1-norm, initial data. This type of analysis is due to DiPerna and Lions [DPa] and develops to obtain existence of weak solutions without smallness assumptions.
- Generalizations of the initial value problems to evolution problems in the presence of boundary conditions.

The relevance of the second class of problems does not need to be emphasized. Indeed, this type of qualitative analysis is developed for general initial conditions (although uniqueness is not obtained) while the first class of results is obtained for very special initial conditions, that often are not related to any interesting physical situation. It is worth mentioning that the third class of problems is particularly relevant for the applications: Real physical systems often include obstacles.

Existence results for the initial value problem solutions were firstly obtained for ***small perturbation from equilibrium*** and, in most cases, make use of the global Maxwellian distribution. That is, they refer to existence, and trend to equilibrium, of the solution of the Cauchy problem that one obtains when the initial conditions are assumed to be small perturbation of Maxwellian equilibrium.

The problem is well documented in the literature, for instance see Chapter 3 of [BLb], the review papers [MSa], [MSb], Chapter 3 of [MSc] and the second and third parts of [WEa]. The spectral theory of the linear operator is reported in the book [KPa]. The book by Greenberg, Protopopescu and van der Mee [GRb] deals with the general theory of linear and linearized operators.

The first study of such a problem is due to Grad [GDa] and refers to local solutions of the equation with hard cutoff potential, starting from a perturbation of Maxwellian. His study was developed in a space domain bounded by specularly reflecting walls. As it is well known, this problem is equivalent to looking for periodic (in the space variable) solutions in the whole space. Grad proves that the linear problem possesses a unique solution linearly growing with time. This result is then used to show ***local*** existence for the nonlinear problem with small distance, in norm, between the initial condition and the Maxwellian.

Later Ukai [UKa] was able to prove that the solution to the linearized problem dealt with by Grad decays to zero as t tends to infinity. This estimate was sufficient to provide the first ***global*** existence proof for initial value problem solutions of the Boltzmann equation in the case, of course, near to equilibrium. Ukai's result was subsequently extended by several authors; in particular, Shizuta proved the existence of classical solutions.

The results we have surveyed above refer to the Boltzmann equation with hard cutoff potential. For the equation with soft cutoff potential the problem was dealt with by Caflisch [CFa], [CFb], who obtained existence and decay to equilibrium in L_2 and L_∞ spaces.

The initial value problem with initial conditions in the whole of $\mathbf{R}^3$ without periodicity conditions involves additional technical difficulties. Local existence was studied by several authors. Then two independent groups: one in the formerly Soviet Union, Maslova and coworkers [MSa]–[MSf], and the other in Japan, Nishida and Imai [NIa], obtained global existence for the equation with hard cutoff. Such a problem was also developed by Ukai [UKb]. Later, essentially at the same time as Caflisch, an important improvement was due to Ukai and Asano [UKc], who proved global existence for the equation with soft cutoff potential in a domain Λ, which may coincide either with the whole space $\mathbf{R}^3$ or with a bounded box with specularly reflecting walls.

A second class of mathematical results refers to the existence and uniqueness of solutions for ***initial conditions close to vacuum***, that is for small initial data which decay to zero at infinity in the phase space. The initial data can be in either $L_1 \cap L_\infty$ or L_∞. In the first case, the mass of the gas is finite. In the second case, the mass can be infinite.

The first proof is due to Illner and Shinbrot [ILa]. Specifically, it refers to a gas of hard spheres decaying exponentially to zero at infinity. Several generalizations have then been produced. For instance, Hamdache [HAa] generalizes the result of [ILa] to the Boltzmann equation with a general pairwise interaction potential. In both papers, the gas is assumed with finite mass. On the other hand, the generalizations produced in [BLa] and [TOa] refer to a gas with infinite mass. In particular, if the initial conditions are assumed to decay, in the physical space, with inverse power behavior then, for sufficiently slow decay, the mass of the gas can indeed be infinite. Both papers refer to the Boltzmann equation with a general pairwise interaction potential.

Further analysis is due to Polewczak [POa] who generalizes to classical solutions the existence theory, previously limited to the case of mild solutions. In [POb], the mathematical results are extended to initial conditions with a more general decay at infinity in the phase space.

Additional developments, somewhat different from those cited above, are due to Shinbrot [SIa] who dealt with the hard spheres Boltzmann equation in the presence of a convex obstacle with deterministically reflecting walls; and to Toscani [TOb] for initial conditions near a local Maxwellian decaying to zero at infinity.

General existence results can be obtained in the functional space $\mathbf{X}(\psi)$ of the functions $f = f(t, \mathbf{x}, \mathbf{v})$ that are μ-measurable with respect to a measure μ finite over the Borel sets of $[0, T] \times \mathbb{R}^3 \times \mathbb{R}^3$, and such that

$$|f^{\#}(t, \mathbf{x}, \mathbf{v})| \leq c\,\psi(\mathbf{x}, \mathbf{v}) \tag{7.4.1}$$

where $f^{\#}$ is defined by (7.3.3), and $\psi(\mathbf{x}, \mathbf{v})$ is a strictly positive bounded continuous function decaying to zero at infinity in the phase space. The functional space $\mathbf{X}(\psi)$ is a Banach space with respect to the norm

$$\|f\| = \operatorname*{ess\,sup}_{t \in [0,T],\ \mathbf{x},\mathbf{v} \in \mathbb{R}^3} \frac{|f^{\#}(t, \mathbf{x}, \mathbf{v})|}{\psi(\mathbf{x}, \mathbf{v})}\,. \tag{7.4.2}$$

The theorem proved by Illner and Shinbrot [ILa] makes use of the so-called Kaniel and Shinbrot iterative scheme [KAa]. Some of the generalizations were obtained by application of classical fixed point theorems.

Another important result on the Cauchy problem refers to ***small perturbation of the spatially homogeneous problem*** and is due to Arkeryd, Esposito, and Pulvirenti [AKc]. Their paper deals with initial data close to the solution of the spatially homogeneous problem, and refers to a gas of particles interacting with a hard cut-off potential and confined to a three-dimensional domain with specularly reflecting boundaries.

The analysis is developed in the following Banach spaces:

- For $s > 1$, let $\mathbf{X}_s$ denote the space of real valued continuous functions $g : \mathbf{v} \in \mathbb{R}^3 \mapsto g(\mathbf{v}) \in \mathbb{R}_+$ such that

$$\|g\|_s = \sup_{\mathbf{v} \in \mathbb{R}^3} (1 + v^2)^s |g(\mathbf{v})| < \infty . \tag{7.4.3}$$

The function $\|\cdot\|_s : \mathbf{X}_s \to \mathbb{R}_+$ represents the norm in $\mathbf{X}_s$.

- For $-\infty < \ell < \infty$, let $\mathbf{H}_\ell$ denote the space of L_2 functions $h : \mathbf{x} \in \Lambda \mapsto h(\mathbf{x}) \in \mathbb{R}_+$ such that

$$|h|_\ell^2 = \sum_{k \in 2\pi \mathbf{Z}^3} (1 + k^2)^\ell |\hat{h}(k)|^2 < \infty , \tag{7.4.4}$$

where $\hat{h}$ denotes the Fourier transform of h, and $\Lambda \subset \mathbb{R}^3$ the spatial domain.

- For $s > 1$ and $-\infty < \ell < \infty$, let $\mathbf{X}_{\ell,s}$ denote the space of Lebesgue measurable functions $b : \Lambda \times \mathbb{R}^3 \to \mathbb{R}$ such that $b(\cdot, \mathbf{v}) \in \mathbf{H}_\ell$ for each $\mathbf{v} \in \mathbb{R}^3$, and $|b(\cdot, \mathbf{v})|_\ell \in \mathbf{X}_s$. The space $\mathbf{X}_{\ell,s}$ is Banach with respect to the norm

$$\|b\|_{\ell,s} = \sup_{\mathbf{v} \in \mathbb{R}^3} (1 + v^2)^s |b(\cdot, \mathbf{v})|_\ell . \tag{7.4.5}$$

The analysis shows that the solution $t \mapsto f(t, \cdot, \cdot)$ of the said problem asymptotically converges in $\mathbf{X}_{\ell,s}$, as t tends to infinity, to the Maxwellian ω with the same mass, momentum, and energy as the initial condition.

A further development, due to Wennberg [WEb], generalizes the existence theorem to the case of pseudo Maxwellian molecules. In addition, it is shown that, for Maxwellian molecules and for particles pairwise interacting with forces harder than the Maxwellian, the above analysis can be developed in spaces more general than $\mathbf{X}_{\ell,s}$. It is proved that in these spaces the solution of the initial value problem strongly converges to the Maxwellian with the same mass, momentum and energy.

7.4.2 The Cauchy problem for large initial data

The analysis of the ***initial value problem for large initial data*** in the L_1-norm is due to DiPerna and Lions [DPa], who proved a theorem which can certainly be classified as the most important one in the analysis of the Cauchy problem in kinetic theory. Their result refers to the existence of weak solutions, according to the definition of ***renormalized solution*** which was given in Section 7.3.

As in the case of existence theory for small initial conditions, one needs suitable assumptions on the collision kernel $\mathcal{B}$:

Assumption 7.4.1. *The collision kernel $\mathcal{B}$ satisfies the condition: $\mathcal{B} \in L_\infty \cap L_1(\mathbb{R}^3; L_1(\mathbf{S}^2))$. Equivalently, that $\mathcal{A}$ belongs to $L_\infty \cap L_1(\mathbb{R}^3)$, where*

$$\mathcal{A}(\mathbf{q}) = \int_{\mathbf{S}^2_+} \mathcal{B}(\mathbf{n}, \mathbf{q})\, d\mathbf{n}\,. \tag{7.4.6}$$

The following lemmas are preliminary to the existence proofs:

Lemma 7.4.1. *Let $f \geq 0$ and $J_i[f]$ belong to $L_{1,loc}([0,\infty) \times \mathbb{R}^3 \times \mathbb{R}^3)$, for $i = 1, 2$. Then f is a solution of Eq. (7.3.2) in the sense of distributions if and only if f is a renormalized solution of (7.3.2). Moreover, if f is a renormalized solution of Eq. (7.3.2) then for each $\beta \in C^1[0,\infty)$ such that*

$$|\beta'(t)|(1+t) \leq c \qquad \text{for some} \quad c \in \mathbb{R}\,, \tag{7.4.7}$$

the composition $\beta(f) := \beta \circ f$ is a solution in the sense of distributions of

$$\left(\frac{\partial}{\partial t} + \langle \mathbf{v}\,, \nabla_{\mathbf{x}} \rangle\right) \beta(f) = \beta'(f)\, J[f]\,, \tag{7.4.8}$$

where the term on the right is well defined in $L_{1,loc}$.

Lemma 7.4.2. *Let $f \geq 0$ and $J_i[f]$ belong to $L_{1,loc}([0,\infty) \times \mathbb{R}^3 \times \mathbb{R}^3)$, for $i = 1, 2$. Then f is a solution of Eq. (7.3.2) in the sense of distributions if and only if it is a mild solution. Moreover,*

i) f is a renormalized solution of Eq. (7.3.2) if and only if it is a mild solution and

$$\frac{1}{1+f}\,J_i[f] \,\in\, L_{1,loc}\,, \qquad i = 1,2\,. \tag{7.4.9}$$

ii) Let

$$F^{\#}(t,\mathbf{x},\mathbf{v}) = \int_0^t B[f]^{\#}(\sigma,\mathbf{x},\mathbf{v})\,d\sigma \tag{7.4.10}$$

where the operator B is defined by, see Eq. (7.2.9),

$$B[f](t,\mathbf{x},\mathbf{v}) := \int\limits_{\Lambda\times\mathbf{S}_+^2} \mathcal{B}(\mathbf{n},\mathbf{q})f(t,\mathbf{x},\mathbf{w})\,d\mathbf{n}\,d\mathbf{w}\,, \tag{7.4.11}$$

and hence: $f\,B[f] \equiv J_2[f]$. Assume that

$$B[f] \in L_1([0,T]\times B_R(0)\times B_R(0))\,, \qquad \forall\, R\,,T < \infty\,, \tag{7.4.12}$$

where $B_R(0)$ is the ball with radius R and center at the origin in $\mathbb{R}^3$. Then f is a mild solution to the initial value problem for Eq. (7.3.2) if and only if the following exponential form is satisfied

$$\begin{aligned} &f^{\#}(t,\mathbf{x},\mathbf{v}) - f^{\#}(s,\mathbf{x},\mathbf{v})\exp\left[-(F^{\#}(t,\mathbf{x},\mathbf{v}) - F^{\#}(s,\mathbf{x},\mathbf{v}))\right] \\ &= \int_s^t J_1[f]^{\#}(\sigma,\mathbf{x},\mathbf{v})\exp\left[-(F^{\#}(t,\mathbf{x},\mathbf{v}) - F^{\#}(\sigma,\mathbf{x},\mathbf{v}))\right]\,d\sigma\,, \\ &\forall\, s,t\,:\, 0 \le s \le t < \infty\,, \quad \text{and a.a. } \mathbf{x},\mathbf{v} \in \mathbb{R}^3\,. \end{aligned} \tag{7.4.13}$$

Formal computations show that any solution of the initial value problem for the Boltzmann equation should also satisfy, for all $t > 0$, the additional equality

$$\int_{\mathbb{R}^6} f(t,\mathbf{x},\mathbf{v})|\mathbf{x}-\mathbf{v}t|^2\,d\mathbf{x}\,d\mathbf{v} = \int_{\mathbb{R}^6} f_0(\mathbf{x},\mathbf{v})|\mathbf{x}|^2\,d\mathbf{x}\,d\mathbf{v}\,. \tag{7.4.14}$$

All this implies that if f is a non-negative solution of (7.3.2), then the following inequality holds

$$\sup_{t\geq 0} \int_{\mathbb{R}^6} f(t,\mathbf{x},\mathbf{v}) \left(1 + |\mathbf{x} - \mathbf{v}t|^2 + v^2 + |\log f(t,\mathbf{x},\mathbf{v})|\right) d\mathbf{x}\, d\mathbf{v} < \infty. \tag{7.4.15}$$

After these preliminaries, the main result given by DiPerna and Lions can be cited in the following two theorems

THEOREM 7.4.1. *Suppose that Assumption 7.4.1 holds true, and that there exists a sequence f_n of non-negative renormalized solutions to the initial value problem for the Boltzmann equation such that:*

$$f_n \in L_\infty\left([0,T] \times \mathbb{R}^3 \times \mathbb{R}^3\right), \qquad \forall T > 0, \tag{7.4.16}$$

$$J_i[f_n] \in L_{1,loc}\left([0,\infty) \times \mathbb{R}^3 \times \mathbb{R}^3\right), \quad i = 1,2, \tag{7.4.17}$$

where f_n satisfies, for some real number C independent of n, the following inequality

$$\sup_{t\geq 0} \int_{\mathbb{R}^6} f_n(t,\mathbf{x},\mathbf{v}) \Big(1 + |\mathbf{x} - \mathbf{v}t|^2 + v^2 + |\log f_n(t,\mathbf{x},\mathbf{v})|\Big)\, d\mathbf{x}\, d\mathbf{v} < C. \tag{7.4.18}$$

Assume that f_n weakly converges in

$$L_1\left([0,T] \times \mathbb{R}^3 \times \mathbb{R}^3\right), \qquad \forall T < \infty,$$

to a function f, and that $f_{n0} = f_n(t = 0)$ weakly converges in $L_1(\mathbb{R}^3 \times \mathbb{R}^3)$ to a function f_0. Then the following properties can be proved:

i) $f \in C\left([0,\infty); L_1(\mathbb{R}^3 \times \mathbb{R}^3)\right)$;

ii) f *satisfies the initial condition* $f(t=0) = f_0$*; moreover, for all* $T < \infty$ *the following holds true*

$$\frac{1}{1+f}\, J_1[f] \quad \in L_1\left([0,T]; L_1(\mathbb{R}^3 \times \mathbb{R}^3)\right), \tag{7.4.19}$$

$$\frac{1}{1+f}\, J_2[f] \quad \in L_\infty\left([0,\infty); L_1(\mathbb{R}^3 \times \mathbb{R}^3)\right), \tag{7.4.20}$$

iii) f *is mild solution, or equivalently a renormalized solution, of the initial value problem for the Boltzmann equation such that*

$$B[f] \;\in\; L_\infty\left([0,\infty); L_1(\mathbb{R}^3 \times \mathbb{R}^3)\right).$$

Moreover, condition (7.4.15) holds true and the ***entropy inequality*** *is satisfied:*

$$H(t) + \frac{1}{4}\int_0^t \int_{\mathbb{R}^6} e(s,\mathbf{x},\mathbf{v})\, ds\, d\mathbf{x}\, d\mathbf{v}$$

$$\leq \liminf_{n\to\infty} \int_{\mathbb{R}^6} f_{n0} \log f_{n0}\, d\mathbf{x}\, d\mathbf{v}, \tag{7.4.21}$$

where H *is defined in Eq. (7.2.12) and where*

$$e(t,\mathbf{x},\mathbf{v}) := \int_{\mathbb{R}^3\times S^2} \mathcal{B}(\mathbf{n},\mathbf{q})\Big(f(t,\mathbf{x},\mathbf{v}')f(t,\mathbf{x},\mathbf{w}')$$

$$- f(t,\mathbf{x},\mathbf{v})f(t,\mathbf{x},\mathbf{w})\Big)\log\left[\frac{f(t,\mathbf{x},\mathbf{v}')f(t,\mathbf{x},\mathbf{w}')}{f(t,\mathbf{x},\mathbf{v})f(t,\mathbf{x},\mathbf{w})}\right] d\mathbf{n}\, d\mathbf{w}. \tag{7.4.22}$$

The main lines of the proof of this theorem are reported in [ARa]. Theorem 7.4.1 can be regarded as a weak stability theorem. In particular, it states that the weak limit of a sequence of classical solutions is a renormalized (or, equivalently, a mild) solution of the Boltzmann equation. From each of them an existence theorem can be stated. In particular

THEOREM 7.4.2. *Under Assumption 7.4.1, suppose that f_0 satisfies $f_0 \geq 0$ almost everywhere in $\mathbb{R}^3 \times \mathbb{R}^3$, and that*

$$\int_{\mathbb{R}^6} f_0(\mathbf{x}, \mathbf{v}) \left(1 + |\mathbf{x}|^2 + v^2 + |\log f_0(\mathbf{x}, \mathbf{v})|\right) d\mathbf{x}\, d\mathbf{v} \; < \; C \qquad (7.4.23)$$

for some $C \in (0, \infty)$. Then there exists an f satisfying items i–iii) of Theorem 7.4.1 and it exists globally in time.

A relevant contribution to the analysis of renormalized solutions is due to P.L. Lions, and is proposed in [LNa]–[LNc] where a remarkable compactness property of the gain term of the Boltzmann equation is proved. Moreover, it is shown that the existence theorem by DiPerna and Lions [DPa] holds true even if the term $|\mathbf{x}|^2$ of Eq.(7.4.23) is replaced by a suitable function $h = h(|\mathbf{x}|)$ satisfying

$$h \geq 0\,, \quad (1 + h)^{\frac{1}{2}} \in \text{Lipschitz}\ (\mathbb{R}^3)\,, \quad \exp(-h) \in L_1(\mathbb{R}^3)\,.$$

Therefore if f_{n0} is the initial condition of the sequence of renormalized solution f_n, then the bound

$$\sup_{n \geq 1} \int_{\mathbb{R}^6} f_{n0}(\mathbf{x}, \mathbf{v}) \left(1 + h(|\mathbf{x}|) + |\mathbf{v}|^2 + |\log f_{n0}(\mathbf{x}, \mathbf{v})|\right) d\mathbf{x}\, d\mathbf{v} < \infty\,,$$

must be satisfied in order to obtain the existence result.

Further, by analyzing only the gain term, the author proves convergence in measure of $J_1[f_n]$ to $J_1[f]$ and its boundedness in suitable L_2-space. This compactness result has several important implications, some of which already indicated by P.L. Lions in the aforementioned papers. We point out the following ones:

i) If f_{n0} converges in $L_1(\mathbb{R}^3 \times \mathbb{R}^3)$ then the sequence f_n converges in $C\left([0, t]; L_1(\mathbb{R}^3 \times \mathbb{R}^3)\right)$ for all finite times.

ii) The solution of the Boltzmann equation for the problem in a periodic box strongly L_1-converges in time, up to the extraction of subsequences, to a pure Maxwellian equilibrium. This result was previously obtained by Arkeryd [AKd] with the methods of nonstandard analysis.

Additional developments are due to Wennberg [WEb], who was able to include the hard-sphere interactions, which were not dealt with in the original proof, and to Mischeler and Perthame [MIa] who considered perturbation of equilibrium for initial conditions with infinite energy.

7.4.3 The initial-boundary value problem

The generalization of the existence theory developed by DiPerna and Lions to the ***initial-boundary value problem*** essentially consists in an analysis that is similar to the one we have just seen for evolution problems, yet includes boundary conditions. The analysis requires several additional developments and sharp inequalities. Limiting this survey to some bibliographical notes, we recall that the first contribution to this topic is due to Hamdache [HAb] for a pointwise reflection law and for a mixture of specular (or reverse) reflection and diffuse reflection. Then, Arkeryd and Cercignani [AKe] studied the case of walls with general diffuse reflection and varying boundary temperature under the restriction of bounded velocities. The problem with unbounded velocities was solved by Arkeryd and Maslova [AKf]. The paper by Goudon and Hamdache [GHa] develops the existence analysis for nonlinear boundary conditions.

Particularly interesting is the analysis developed by Arkeryd and Nouri [AKg] about the asymptotics of the initial boundary value problem and the strong L_1 convergence analysis towards equilibrium.

7.4.4 Open problems

The problems of existence theory for ***large initial data with uniqueness*** and of finding constructive methods to obtain quantitative information on the solution are still open. This aspect is particularly important in view of the generalized Boltzmann models. However, the solution of this problem seems quite remote. On the other hand, several existence and uniqueness results can be obtained for modified Boltzmann models. This aspect is not so interesting if the modifications are artificially developed only to reach stronger results and without the due support from physics, but sometimes the modifications may in fact be fully motivated on the level of physics.

With this in mind, we recall that the Boltzmann equation was derived under the assumption, among others, that the two particles that are going to collide are uncorrelated and that the variations of f are negligible at distances of the order of the radius R of the effective cross sectional area. The latter assumption implies that the spatial argument in the one-particle distribution functions is $\mathbf{x} = \mathbf{x}_1 = \mathbf{x}_2$, where $\mathbf{x}_1$ and $\mathbf{x}_2$ are the centers of the two colliding particles.

In order to obtain a more detailed description of physical reality one should compute the trajectories of the particles during the interaction, and estimate the evolution of the distribution function along the trajectories. This type of modelling leads to the generalized Boltzmann equation, which involves somewhat complicated calculations and can be regarded as a well understood model only in the spatially homogeneous case.

An indirect way to model the kinetic equation that avoids the above stated simplification consists in averaging the distribution function of the field particles over the action domain of the test particles. Examples of this type of modelling are: the one reported in [BLe] and those proposed in the papers by Morgenstern [MGa] and Povzner [PVa]. Indeed, if the distribution function is averaged over a domain in the phase space, then the evolution operator U assumes all the properties needed to develop a nice existence and uniqueness theory for the initial value problem. The crucial point is to define physically consistent averages instead of purely artificial manipulation meant to simplify the mathematical problem.

The interesting aspect of the averaging proposed in [BLe] is that it contains, as limit cases, both the Boltzmann equation and the Enskog equation. In particular the Enskog model, taking into account particles of finite size, is a candidate for modelling moderately dense gases. A detailed description of Enskog original model, together with various developments and relevant results, are reported in [BLd].

The analysis of the initial-boundary value problem for averaged models of the Boltzmann equation requires a setting of the boundary conditions somehow different from the one stated in Eq. (7.3.11). In fact, interactions should not be local in space; instead, averages should be performed over the action domain of the particle hitting the wall. The properties of the operator Q, introduced in Eq. (7.3.11), should also be revisited. This topic, however, is not yet dealt with in the literature.

7.4.5 Evolution problems in the presence of a force field

The mathematical results we have just reviewed refer to evolution problems in the absence of an external force field. A first generalization of the above result is the analysis of problems in the ***presence of a force field***. The simplest case is when the force does not depend on the distribution function.

The analysis of problems for small initial conditions follows methods similar to those reviewed in subsection 7.4.1. The paper by Asano [ASa] provides one of the first relevant results on this topic. He proves local existence for quite general initial conditions. Asano's formulation is the starting point for all the studies later developed on this subject, see [ASb], [GUa], [BLc]. In particular, the papers by Asano [ASb] and Grunfeld [GUa] provide a global existence proof for the solution with initial conditions close to equilibrium, and a conservative force field driving the system towards equilibrium. In [BLd] the mathematical theory for decaying data is generalized to the equation in the presence of a force field.

More interesting is the analysis in [LNc] where, among various results, existence is proved for the equation in the presence of Vlasov-like forces. Boundary value problems in the presence of a force field are not yet extensively dealt with despite the great interest of this type of problems for the applications.

7.4.6 Shock waves

The ***shock wave problem*** was described in Eqs. (7.3.16) as a boundary value problem in one space dimension. The main result on this topic, before DiPerna and Lions theory, was proposed by Caflisch and Nicolaenko [CFc]. Their analysis is developed in the framework of the hydrodynamic description of kinetic theory. In particular, it is proved that for sufficiently closed Maxwellians (weak shock waves) the Boltzmann equation describes a shock profile that is close, in a suitable norm, to that described by the Navier Stokes model.

The smallness assumption is no longer necessary in the analysis developed by Lions [LNd], where the introduction of the concept of relative entropy allows to generalize his compactness analysis and obtain distributional solutions.

The problem can also be studied in terms of moment approximation, as documented in the book by Cercignani [CEa], or by application of numerical schemes as shown in the paper by Okwade [OKa]. In particular, in this nice paper several detailed computations are presented and further analysis is motivated in the direction of existence and uniqueness of strong shock wave profiles.

7.4.7 Asymptotic analysis

The ***asymptotic analysis towards hydrodynamic limits***, and the mathematical problems related to the ***derivation of the Boltzmann equation***, are the last two topics that complete the analysis of this model. For both subjects only a brief description and some bibliographical indications will be given.

The first topic is documented, among several others, in [BGa], [BGb], [DMa] and [LAa], [LAb]. The survey [LAc] also provides an interesting unified treatment of the various approaches to this difficult mathematical problem with reference not only to the Boltzmann equation, but also to the Enskog model and to averaged models such as those described in [BLf].

The analysis consists in developing a perturbation expansion of the model when written in suitable dimensionless form with respect to its small parameter. After a detailed analysis about the existence and integrability properties of the solution, several hydrodynamic equations can be obtained, one for each term of the power expansion. Different hydrodynamic limits correspond to different scaling models for the kinetic equation.

The phenomenological derivation of Boltzmann equation was obtained, as we have seen, under assumptions that cannot be justified on a mathematical basis. In particular, the factorization of the two-particle distribution function, that is necessary to close the BBGKY hierarchy, is certainly false for general initial and boundary conditions. Actually, this assumption may be true only for very special conditions, related to the analysis of a rarefied gas in vacuum. Still several problems remain open. Some of them are: the link between distributional solutions of the Boltzmann equation and the model derivation, the analysis of models such as the Enskog one, the analysis of the averaged models.

7.5 Computational Methods

Solving mathematical problems related to the Boltzmann equation, that is solving the initial-boundary value and boundary value problems, requires overcoming several difficulties related to the intrinsic complexity of the equation.

The dependent variable of the evolution equation is a function of time, space and velocity. In the absence of an external force field, the left-hand side may be computed only if discretization of the time and space variables is performed, and time and space derivatives consequently approximated. Moreover, owing to the presence of the space variable, to estimate the collision integral one has to discretize the velocity as well, even if it acts as a parameter.

In principle, by discretizing the space, velocity, and angular variables (i.e., the vector $\mathbf{n}$), one produces a number of (ordinary differential) equations given by the number of space variable collocation points multiplied by the numbers of collocation points relative to the velocity and to the angular variables; hence the system one obtains is practically impossible to be solved. In addition, if a force field is present the difficulty increases enormously since in this case the velocity becomes a further independent variable.

Alternative methods more powerful than the traditional ones are needed. In particular when dealing with practical problems, both external and internal flows may require the splitting of physical space into two, possibly overlapped, domains: one for the Boltzmann equation to be valid, the other for the continuum mechanics equation. In this way computations are reduced since they are much lighter for the second equation than they are for the first one.

In details, and with reference to [TIa], [BTa] and to the survey [BLg], various steps of the methods are the following

- Decomposition of the external field domain into the (possibly overlapped) domains of validity respectively of the Boltzmann's and of the hydrodynamics equation;
- Solution of the kinetic equation in the domain of validity of the Boltzmann equation;
- Solution of the hydrodynamic equation in their domain of validity;
- Coupling of the continuous and kinetic solutions of the two models.

This procedure is typical of the applications of the Boltzmann equation to the analysis of fluid dynamical problems and is summarized in the second part of the review paper [BLg], where a detailed information can be recovered on this topic and the related literature. Considering that we are interested in generalized Boltzmann models, our attention will be limited to the solutions of the equation. Therefore we do not report in this chapter any of the problems concerning either the decomposition or the coupling problems.

Conversely, in presenting the content of this section we will take advantage of the review by Walus [WAa]. An even more complete review, also containing several interesting applications, is the one by Neunzert and Struckmeier [NEb]; additional information can also be recovered in [GNa], [GNb] and [NEa].

The topic which is more closely related to the contents of this book, is the ***computational*** analysis of the kinetic equations, hence one has to tackle the problem of reducing the number of collocation points in the phase space. For nonstationary problems the major approaches are based on the ***splitting method***, a procedure which (in the case of deterministic methods) decouples at each time step the Boltzmann equation into two equations: the spatially homogeneous Boltzmann's and the collisionless transport equation; both to be then solved numerically by the appropriate schemes. The method was first applied by Temam [TEa] to several evolution problems and, in particular, to the Broadwell model. Here we shall give a brief description of this procedure with reference to the survey paper [WAa], where additional information can be recovered.

Suppose that the time interval $[0, T]$ has been divided into N equal subintervals of length $\tau = T/N$, and suppose that on the time interval $((k-1)\tau, k\tau]$ the (constant) value f^k has been computed for the approximate distribution function f_τ, where $k = 1, \ldots, N-1$. Then the value f^{k+1} of the distribution function on the next time interval $(k\tau, (k+1)\tau]$ is obtained in two steps.

In the first step, one solves the ***spatially uniform relaxation problem***

$$\frac{\partial f^*}{\partial t} = J[f^*] \quad \text{on} \quad (k\tau, (k+1)\tau]\,, \qquad f^*(k\tau) = f^k\,. \tag{7.5.1}$$

The solution of (7.5.1) evaluated at the endpoint $(k+1)\tau$ serves then as the initial value for the second step, which corresponds to the ***free flow problem***

$$\begin{cases} \dfrac{\partial f^{**}}{\partial t} + \langle \mathbf{v}\,, \nabla_{\mathbf{x}} \rangle f^{**} = 0 \quad \text{on} \quad (k\tau, (k+1)\tau]\,, \\ f^{**}(k\tau) = f^*((k+1)\tau)\,. \end{cases} \tag{7.5.2}$$

Finally

$$f^{k+1} = f^{**}((k+1)\tau)\,. \tag{7.5.3}$$

In the deterministic framework, solution means further discretizing the position and velocity variables and solving the algebraic system relative to the distribution function values at the grid points. In the stochastic framework, solution means covering the spatial domain with cells and simulating a stochastic process for a finite number of molecules distributed over the cells [AVa]. Moreover, in the case of an initial boundary value problem, appropriate boundary conditions are to be taken into account when solving the free flow stage.

The convergence of the procedure was established by Bogomolov [BOa]. Recently, a convergence of the splitting procedure towards the DiPerna-Lions solutions of the Boltzmann has been investigated by Desvillettes and Mischeler [DEa].

Since the free flow equation is generally solved numerically by use of standard finite difference and volume schemes, its crucial point being the preservation of the conserved quantities, it is important to carefully deal with the first step of the procedure, that is in a correct evaluation of the collision operator.

The first step is to approximate the local distribution function f on a suitable set of cells C_j. After the pioneer work by Bird (also reported in his books [BRa] and [BRb]), the first algorithm directly related to the Boltzmann equation seems to be due to Nanbu [NAa]-[NAc]; then, some developments and related convergence analysis are due to Babovsky [BAa]-[BAc] and Babovsky-Illner [BAe]; researches towards applications can be found in [GNa], [GNb], [ILb], [NEa]; convergence proofs are developed, among others, in [PUa] and [WGa]; an overall description of the whole numerical process with several interesting applications may be found in the surveys [NEa] and [NEb].

As an alternative to the above methods, the regular quadratures method for the collision operator may be used. It can be developed for distribution functions which have been suitably approximated, or interpolated, over suitable collocation points in the velocity space. Generally these approximations lead to a discretized Boltzmann equation which resembles the equations for discrete-velocity models in kinetic theory. This approach has been recently applied by several authors: Preziosi and Longo [PRa], Preziosi and Rondoni [PRb], Buet [BUa] and [BUb], Bobylev et al. [BPa]. In particular, Preziosi operates in polar coordinates, and this has the great advantage that only the velocity modulus is defined on an infinite domain; discretization of the velocity variable is then developed in such a way that the collision operator still retains all the conservation properties of the original Boltzmann equation.

7.6 References

[AKa] ARKERYD L., Asymptotic behavior of the Boltzmann equation with infinite range force, *Comm. Math. Phys.*, **86** (1982), 475–484.

[AKb] ARKERYD L., On the Boltzmann equation in unbounded space far from equilibrium, *Comm. Math. Phys.*, **105** (1982), 205–219.

[AKc] ARKERYD L., ESPOSITO M., and PULVIRENTI M., The Boltzmann equation for weakly inhomogeneous data, *Comm. Math. Phys.*, **111** (1987), 393–407.

[AKd] ARKERYD L., On the strong L_1 trend to equilibrium for the Boltzmann equation, *Stud. Appl. Math.*, **87** (1992), 282–293.

[AKe] ARKERYD L. and CERCIGNANI C., A global existence theorem for the initial-boundary value problem for the Boltzmann equation when boundaries are not isothermal, *Arch. Rat. Mech. Anal.*, **125** (1993), 271–287.

[AKf] ARKERYD L. and MASLOVA N., On diffuse reflection at the boundary for the Boltzmann equation and related equations, *J. Stat. Phys.*, **77**, (1994), 1051–1077.

[AKg] ARKERYD L. and MASLOVA N., Asymptotics of the Boltzmann equation with diffuse reflection boundary conditions, *Rapporte Interne Université de Nice*, n. 406, (1994).

[ARa] ARLOTTI L. and BELLOMO N., On the Cauchy problem for the nonlinear Boltzmann equation, in **Lecture Notes on the Mathematical Theory of the Boltzmann Equation**, Bellomo N. Ed., World Scientific, London, Singapore, (1995).

[ASa] ASANO K., Local solutions to the initial and initial boundary value problem for the Boltzmann equation with an external force, *J. Math. Kyoto University*, **24** (1984), 225–238.

[ASb] ASANO K., Global solutions to the initial boundary value problem for the Boltzmann equation with an external force, *Transp. Theory Stat. Phys.*, **16** (1987), 735–761.

[AVa] ARSEN'EV A.A., On the approximation of the Boltzmann equation by the stochastic differential equations, *Zh. Vychisl. Mat. Mat. Fiz.*, **28** (1988), 500–567 (in Russian).

[BAa] BABOVSKY H., On a simulation scheme for the Boltzmann equation, *Math. Methods Appl. Sci.*, **8** (1986), 223–233.

[BAb] BABOVSKY H., A convergence proof for Nanbu's Boltzmann simulation scheme, *Eur. J. Mech. B Fluids*, **8** (1989), 45–55.

[BAc] BABOVSKY H., Monte Carlo simulation schemes for steady kinetic equations, *Transp. Theory Stat. Phys.*, **23**(1-3) (1994), 249–264.

[BAd] BABOVSKY H., GROPENGIESSER F., NEUNZERT H., STRUCKMEIER J., and WEISEN B., Application of well-distributed sequences to the numerical simulation of the Boltzmann equation, *J. Comp. Appl. Math.*, **31** (1990), 15–22.

[BAe] BABOVSKY H. and ILLNER R., A convergence proof for Nanbu simulation method for the full Boltzmann equation, *SIAM J. Numer. Anal.*, **26** (1989), 45–65.

[BGa] BARDOS C., GOLSE F., and LEVERMORE D., Fluid dynamics limits of kinetic equations I: Formal derivation, *J. Stat. Phys.*,

63 (1991), 323–344.

[BGb] BARDOS C., GOLSE F., and LEVERMORE D., Fluid dynamics limits of kinetic equations II, *Comm. Pure Appl. Math.*, **46** (1993), 235–273.

[BLa] BELLOMO N. and TOSCANI G., On the Cauchy problem for the nonlinear Boltzmann equation: Global existence, uniqueness and asymptotic behavior, *J. Math. Phys.*, **26** (1985), 334–338.

[BLb] BELLOMO N., PALCZEWSKI A., and TOSCANI G., **Mathematical Topics in Nonlinear Kinetic Theory**, World Scientific, London, Singapore, (1988).

[BLc] BELLOMO N., LACHOWICZ M., PALCZEWSKI A., and TOSCANI G., On the initial value problem for the Boltzmann equation with force term, *Transp. Theory Stat. Phys.*, **18** (1989), 87–102.

[BLd] BELLOMO N., LACHOWICZ M., POLEWCZAK J., and TOSCANI G., **Mathematical Topics in Nonlinear Kinetic Theory II: The Enskog Equation**, World Scientific, London, Singapore, (1991).

[BLe] BELLOMO N. and POLEWCZAK J., The Generalized Boltzmann Equation: Solution and exponential convergence to equilibrium, *Transp. Theory Stat. Phys.*, **26** (1997), 661–677.

[BLf] BELLOMO N. Ed., **Lecture Notes on the Mathematical Theory of the Boltzmann Equation**, World Scientific, London, Singapore, (1995).

[BLg] BELLOMO N., LE TALLEC P., and PERTHAME B., On the Solution of the Nonlinear Boltzmann Equation, *ASME Review*, **48** (1995), 777–794.

[BOa] BOGOMOLOV S.V., The convergence of the splitting method for the Boltzmann equation, *Zh. Vychisl. Mat. Mat. Fiz.*, **28** (1988), 119–126 (in Russian).

[BPa] BOBYLEV A.V., PALCZEWSKI A., and SCHNEIDER J., Approximation of the Boltzmann discrete velocity model, *Comp. Rend. Acad. Sci. Paris*, **320** (1995), 367–370.

[BRa] BIRD G.A., **Molecular Gas Dynamics**, Oxford University Press, Oxford, (1976).

[BRb] BIRD G.A., **Molecular Gas Dynamics and the Direct Simulation of Gas Flows**, Clarendon Press, New York, (1994).

[BTa] BOURGAT J.F., LE TALLEC P., TIDRIRI D., and QIU Y., Numerical coupling of nonconservative or kinetic models with the conservative compressible Navier–Stokes equations, INRIA Report n. 1426 (1991).

[BUa] BUET C., A discrete-velocity scheme for the Boltzmann operator of rarefied gas dynamics, *Transp. Theory Stat. Phys.*, **24** (1995).

[BUb] BUET C., Conservative and entropy schemes for the Boltzmann collision operator of polyatomic gases, *Math. Models Meth. Appl. Sci.*, **7** (1998), 165–192.

[CEa] CERCIGNANI C., **Theory and Application of the Boltzmann Equation**, Springer, Heidelberg, (1988).

[CEb] CERCIGNANI C., ILLNER R. and PULVIRENTI M., **Theory and Application of the Boltzmann Equation**, Springer, Heidelberg, (1993).

[CEc] CERCIGNANI C., LAMPIS M., and LENTATI A., A new scattering kernel in kinetic theory of gases, *Transp. Theory Stat. Phys.*, **24** (1995), 1319–1336.

[CFa] CAFLISCH R.E., The Boltzmann equation with a soft potential: Part I, *Comm. Math. Phys.*, **74** (1980), 71–95.

[CFb] CAFLISCH R.E., The Boltzmann equation with a soft potential: Part II, *Comm. Math. Phys.*, **74** (1980), 97–59.

[CFc] CAFLISCH R.E. and NICOLAENKO B., Shock wave solution

of the Boltzmann equation, *Comm. Math. Phys.*, **74** (1980), 71–95.

[CMa] CHORIN A. and MARSDEN J., **A Mathematical Introduction to Fluid Dynamics**, Springer, Heidelberg, (1979).

[DEa] DESVILLETTES L. and MISCHELER S., About the splitting algorithm for Boltzmann and B.G.K equations, *Math. Models Meth. Appl. Sci.*, **8** (1996), 1079–1102.

[DMa] DE MASI A., ESPOSITO R., and LEBOWITZ J., Incompressible Navier-Stokes and Euler limits of the Boltzmann equation, *J. Statist. Phys.*, **42** (1989), 1189–1214.

[DPa] DIPERNA R. and LIONS P.L., On the Cauchy problem for the Boltzmann equation: Global existence and weak stability results, *Annals Math.*, **130** (1990), 1189–1214.

[DPb] DIPERNA R. and LIONS P.L., Global solutions of the Boltzmann and the entropy inequality, *Arch. Rat. Mech. Anal.*, **114** (1991), 47–55.

[GDa] GRAD H., Asymptotic equivalence of the Navier–Stokes and the nonlinear Boltzmann equation, in **Proc. Symp. Appl. Math.**, American Math. Soc. Publ., **17** (1965), 154–183.

[GLa] GLASSEY R., **The Cauchy Problem in Kinetic Theory**, SIAM Publ., Philadelphia, (1995).

[GNa] GROPENGIESSER F., NEUNZERT H., STRUCKMEIER J., and WIESEN B., Hypersonic flow around a 3D–Delta wing at low Knudsen numbers, in **Rarefied Gas Dynamics**, Beylich A. Ed., VCH Press (1990), 332–336.

[GNb] GROPENGIESSER F., NEUNZERT N., and STRUCKMEIER J., Computational methods for the Boltzmann equation, in **Applied and Industrial Mathematics**, Spigler R. Ed., Kluwer Academic Publisher, Amsterdam, (1991), 19–49.

[GRa] GREENBERG W., ZWEIFEL P., and POLEWCZAK J., Global existence proofs for the Boltzmann equation, in **Nonlinear**

Phenomena: The Boltzmann Equation, Lebowitz J. and Montroll E. Eds., North-Holland, Amsterdam, (1983), 21–49.

[GRb] GREENBERG W., VAN DER MEE C.V.M., and PROTOPOPESCU V., **Boundary Value Problems in Abstract Kinetic Theory**, Birkhäuser, Basel, (1987).

[GUa] GRUNFELD C.P., On the nonlinear Boltzmann equation with force term, *Transp. Theory Stat. Phys.*, **14** (1985), 291–322.

[HAa] HAMDACHE K., Existence in the large and asymptotic behavior for the Boltzmann equation, *Japan J. Appl. Math.*, **2** (1985), 1–15.

[HAb] HAMDACHE K., Initial-boundary value problems for the Boltzmann equation: Global existence of weak solutions, *Arch. Rat. Mech. Anal.*, **119** (1992), 309–353.

[ILa] ILLNER R. and SHINBROT M., The Boltzmann equation: Global existence for a rare gas in an infinite vacuum, *Comm. Math. Phys.*, **95** (1984), 117–126.

[ILb] ILLNER R. and NEUNZERT H., On simulation methods for the Boltzmann equation, *Transp. Theory Stat. Phys.*, **16** (1987), 141–154.

[KAa] KANIEL S. and SHINBROT M., The Boltzmann equation: uniqueness and local existence, *Comm. Math. Phys.*, **58** (1984), 65–84.

[KPa] KAPER H., LEKKERKERKER C. and HEJTMANEK J., **Spectral Theory in Linear Transport Equation**, Birkhäuser, Basel, (1982).

[LAa] LACHOWICZ M., On the initial layer and the existence theorem for the nonlinear Boltzmann equation, *Math. Methods Appl. Sci.*, **9** (1987), 342–366.

[LAb] LACHOWICZ M., On a system of stochastic differential equations the initial layer and the existence theorem modelling the Euler and the Navier-Stokes equations, *Japan J. Appl. Math.*, **10** (1993), 109–131.

[LAc] LACHOWICZ M., Asymptotic analysis of nonlinear kinetic equations: The hydrodynamic limit, in **Lecture Notes on the Mathematical Theory of the Boltzmann Equation** , Bellomo N. Ed., World Scientific, London, Singapore, (1995).

[LNa] LIONS P.L., Compactness in Boltzmann equation via Fourier integrals operators and applications I, *J. Math. Kyoto University*, **34** (1994), 391–427.

[LNb] LIONS P.L., Compactness in Boltzmann equation via Fourier integrals operator and applications II, *J. Math. Kyoto University*, **34** (1994), 429-460.

[LNc] LIONS P.L., Compactness in Boltzmann equation via Fourier integrals operator and applications III, *J. Math. Kyoto University*, **34** (1994), 539–584.

[LNc] LIONS P.L., Conditions at infinity for Boltzmann equation, *Comm. Partial Diff. Equations*, **19** (1994), 335–367.

[MGa] MORGENSTERN D., Analytic studies related to the Maxwell–Boltzmann equation, *Arch. Rat. Mech. Anal.*, **4** (1955), 533-555.

[MIa] MISCHLER S., and PERTHAME B., Boltzmann equation with infinite energy: Renormalized solutions and distributional solutions for small initial data and initial data close to a Maxwellian, *SIAM J. Math. Anal.*, **28** (1997), 1015–1027.

[MSa] MASLOVA N., Existence and uniqueness theorems for the Boltzmann equation, in **Dynamical Systems II**, Sinai Ya.G. Ed., Springer, Heidelberg, (1992), 254–279.

[MSb] MASLOVA N., Mathematical methods of studying the Boltzmann equation, *St. Petersburg Math. J.*, **1** (1992), 41–80.

[MSc] MASLOVA N., **Nonlinear Evolution Equations**, World Scientific, London, Singapore, (1993).

[MSd] MASLOVA N. and FIRSOV A.N., Solutions of the Cauchy Problem for the Boltzmann equation, *Vest. Leningrad*, **19** (1975), 83–88, (in Russian).

[MSe] MASLOVA N. and TCHUBENKO R.P., On the solutions of the nonstationary Boltzmann equation, *Dokl. Akad. Nauka USSR*, **202** (1972), 800–803, (in Russian).

[MSf] MASLOVA N. and TCHUBENKO R.P., Asymptotic properties of the solutions of the Boltzmann equation, *Vest. Leningrad*, **1** (1973), 100–105, (in Russian).

[NAa] NANBU K., Direct simulation scheme derived from the Boltzmann equation. I. Monocomponent gases, *J. Phys. Soc. Japan*, **49**, No. 5 (1980), 2042–2049.

[NAb] NANBU K., Stochastic solution method of the Master Equation and the model Boltzmann equation, *J. Phys. Soc. Japan*, **52**, No. 8 (1983), 2654–2658.

[NAc] NANBU K., Derivation from Kac's Master equation of the stochastic laws for simulating molecular collisions, *J. Phys. Soc. Japan*, **52**, No. 12 (1983), 4160–4165.

[NEa] NEUNZERT H., Modelling and numerical simulation of collisions, *Int. Rep. n. 96, Arbeit. Technomathematik*, (1993).

[NEb] NEUNZERT H. and STRUCKMEIER J., Particle Methods for the Boltzmann equation, in **Acta Numerica 1995**, Cambridge University Press, (1995), 417–458.

[NIa] NISHIDA T. and IMAI K., Global solutions to the initial value problem for the nonlinear Boltzmann equation, *Publ. RIMS Kyoto University*, **12** (1976), 229–239.

[POa] POLEWCZAK J., Classical solution of the nonlinear Boltzmann equation in all space: Asymptotic behavior of the solutions, *J. Stat. Phys.*, **50** (1988), 611–632.

[POb] POLEWCZAK J., New estimates of the nonlinear Boltzmann operator and their application to existence theorem, *Transp. Theory Stat. Phys.*, **18** (1988), 235–247.

[PRa] PREZIOSI L., and LONGO E., On a conservative polar discretization of the Boltzmann equation, *Japan J. Ind. Appl. Math.*, **14** (1997), 399–435.

[PRb] PREZIOSI L., and RONDONI L., Conservative energy discretization of the Boltzmann collision operator, *Quarterly J. Appl. Math.*, to appear.

[PVa] POVZNER A.Y., The Boltzmann equation in the kinetic theory of gases, *American Math. Soc. Transl. Ser. 2*, **47** (1962), 193–216.

[PUa] PULVIRENTI M., WAGNER W., and ZAVELANI ROSSI M., Convergence of particle schemes for the Boltzmann equation, *Eur. J. Mech. B/Fluids*, **13** (1994), 339–351.

[SHa] SHIZUTA Y., On the classical solution of the Boltzmann equation *Comm. Pure Appl. Math.*, **36** (1983), 705–754.

[SIa] SHINBROT M., Flow of a cloud past a body, *Transp. Theory Stat. Phys.*, **15** (1986), 317–332.

[TEa] TEMAM R., Sur la stabilité et la convergence de la mèthode des pas fractionaires, *Ann. Mat. Pura Appl.*, **79** (1968), 191–380.

[TIa] TIDRIRI D., Couplage d'approximations et de modèles de type différents dans le calcul d'écoulements externes, Thèse, Paris 9, Mai 1992.

[TOa] TOSCANI G., On the Boltzmann equation in unbounded domains, *Arch. Rat. Mech. Anal.*, **95** (1986), 37–49.

[TOb] TOSCANI G., Global solution of the initial value problem for the Boltzmann equation near a local Maxwellian, *Arch. Rat. Mech. Anal.*, **102** (1988), 231–241.

[TRa] TRUESDELL C. and MUNCASTER R., **Fundamentals of Maxwell Kinetic Theory of a Simple Monoatomic Gas**, Academic Press, New York, (1980).

[UKa] UKAI S., On the existence of global solutions of mixed problem for the nonlinear Boltzmann equation, *Proc. Jap. Acad.*, **50** (1974), 179–184.

[UKb] UKAI S., Les solutions globales de l'équation non–linéaire de

Boltzmann dans l'espace tout entier et le demispace, *Comp. Rend. Acad. Sci. Paris*, **A282** (1976), 317–320.

[UKc] UKAI S. and ASANO K., On Cauchy problem for the nonlinear Boltzmann equation with soft potentials, *Publ. RIMS Kyoto University*, **18** (1982), 477–519.

[WAa] WALUS W., Current computational methods for the nonlinear Boltzmann equation, in **Lecture Notes on the Mathematical Theory of the Boltzmann Equation**, Bellomo N. Ed., World Scientific, London, Singapore, (1995).

[WEa] WENNBERG B., **Stability and Exponential Convergence for the Boltzmann Equation**, Thesis, Dept. of Math., Göteborg and Chalmers University of Technology (1993).

[WEb] WENNBERG B., Regularity estimates for the Boltzmann equation, Internal report No. 1994 - 02, Dept. Math. University Göteborg (1994).

[WGa] WAGNER W., A convergence proof for Bird's direct simulation Monte Carlo method for the Boltzmann equation, *J. Stat. Phys.*, **66** (1992), 1011–1044.

Chapter 8

Generalized Kinetic Models for Traffic Flow

8.1 Introduction

This chapter is devoted to the modelling of one-dimensional flows of vehicles along a road in the particular framework of the kinetic approach. Various analyses of this subject can be found in the literature; basically, they may be separated in three different classes. Models which belong to the first group have either a dynamical system structure or a cellular automata structure. Models of the second group are essentially fluid-dynamical. Models of the third kind are developed on a statistical basis, and may be considered as a connection between those of the first two groups.

More in details, the three different lines are:

- Microscopic ***follow-the-leader*** or ***cellular automata*** models (see for instance [GAa] or [NAa], [NSa]), which consist either on a set of second order ordinary differential equations, or on finite difference mappings. Each equation describes the dynamics of a single vehicle in a framework close to Newtonian mechanics. According to this picture, the behavior of each vehicle is seen as a direct consequence of the preceding one.
- Macroscopic ***hydrodynamic*** models which heuristically derive suitable evolution equations for the density and momentum of the fluid of vehicles. They are based on the continuum hypothesis, on

some convenient conservation equations, and on phenomenological relations.

- Mesoscopic ***kinetic-type*** models which deal with the statistical number-of-car distribution function and its time, space, and velocity dependence. In this case, the mathematical model consists in a nonlinear integrodifferential evolution equation of the Boltzmann type based on the interactions that, at the microscopic scale, occur among the vehicles.

According to the general subject of this book, we shall only develop the third one of these lines. However, to better understand the difficulties and advantages of this framework, and to acquire new perspectives and hints on this research field, a brief introduction to the hydrodynamic approach may be of some help. Indeed, the macroscopic picture is, ultimately, the reference framework within which measurements are actually performed and experimental validations possible.

The content of the chapter is organized into four sections.

Section 8.1 is this introduction.

Section 8.2 deals with the hydrodynamic approach.

Section 8.3 develops the analysis of interactions between pairs of vehicles, and shows some possible derivations of the evolution equations for the vehicles distribution function.

Section 8.4 contains a critical analysis on the matter, and provides indications for further research perspectives.

8.2 Traffic Flow and Hydrodynamics

A considerable amount of work has been done in the field of macroscopic description of traffic flows. Research activity in this field is documented in specialized books such as the one by Leutzback [LEa]. A concise report on the state of the art can also be recovered in the papers by Klar and Wegener [KLa]–[KLc].

Here we shall only recall some of the main results of this modelling, and point out their characteristics. In particular we begin with the following introductory discussion.

On the one hand, it is undeniable that most of the observations on vehicular traffic concern macroscopic variables such as mass velocity, flow, density, pressure, etc. Experimental measurements relative to vehicular traffic are designed to acquire the functional dependencies among these variables. For instance: the ***fundamental diagram*** reports about the behavior of mean speed versus density. Moreover, observations often rely on an additional hypothesis: the equilibrium assumption. This *a priori* condition is suggested, and convenient, especially when one is faced with the actual difficulties of performing sensible and accurate measurements.

On the other hand, from the modelling point of view the continuum hypothesis is not easily sustainable. Therefore more detailed descriptions are needed in spite of the computational impact of these models that may be much higher. This motivates the various developments of kinetic models for traffic flow.

With this in mind, and taking care of avoiding unmotivated similarities, as long as the hydrodynamic framework is assumed to be true the classical variables and procedures are of use, even if they are completely justified only in the real physical world of hydrodynamic. When necessary, for instance to close the procedure that recovers macroscopic equations starting from the kinetic ones, they may be added by some phenomenological relations based on the observations of real traffic equilibrium dynamics.

Once the model is proposed, and simulations obtained by solving the related initial-boundary value problems, validations may be accomplished only by referring to the phenomenologic dependencies that have actually been observed in connection with the macroscopic variables.

Unluckily, repeatability of experiments on general flow conditions appears impossible to organize. Therefore, as in nonequilibrium thermodynamics, not only the macroscopic variables such as the vehicle density and mass velocity are computed by averaging the corresponding local variables, but also the model validations are achieved by recourse to effective observations only at equilibrium.

Starting point of a macroscopic model are the mass and momentum conservation equations; yet the procedure may be generalized to include even other variables and equations. This shows how to subdivide the hydrodynamic models: scalar models and vector models. According to both of them, the mass density is a dependent variable of time and space, and both of them require the mass conservation equation. The two classes differ in that the vector models associate a second variable to the density, namely: the mass velocity; and require another equation: the momentum conservation. Hence, the latter is an independent and further condition that is assumed to hold on the system in relation with the acceleration procedures. In the scalar models instead, on the basis of phenomenological observations, the velocity variable is *a priori* assumed to functionally depend on the density itself and on the density gradient.

To be more precise, let us introduce the following notation that describes a one-dimensional continuum fluid of vehicles along a road of length ℓ during a time period τ. We shall call

$$N = N(s,z) : [0,\tau] \times [0,\ell] \to \mathbb{R}^+ \,, \tag{8.2.1}$$

and

$$V = V(s,z) : [0,\tau] \times [0,\ell] \to \mathbb{R}^+ \,, \tag{8.2.2}$$

respectively the ***density*** (number of vehicles per unit tract) and the (mass) ***velocity*** of the fluid of vehicles, at time s and point z. The classic, e.g., [CMa], pair of coupled equations is: one for the mass balance equation, the other for momentum

$$\begin{cases} \dfrac{\partial N}{\partial s} + \dfrac{\partial}{\partial z}(NV) = 0\,, \\[2ex] \dfrac{\partial V}{\partial s} + V\dfrac{\partial V}{\partial z} = \widetilde{F}\,(N, V, \ldots)\,, \end{cases} \tag{8.2.3}$$

where $\widetilde{F}$ is the force density applied to the vehicles. For notation sake, the following independent and dependent variables are introduced, conveniently rescaled so that they take values on the interval $[0, 1]$. We shall call

$x = z/\ell$ the dimensionless space variable referred to the length of the road;

$t = s/\tau$ the dimensionless time variable referred to the observation time τ. As it will be seen, the time τ may be suitably chosen.

$n\ (t, x) = N(\tau t, \ell x)/N_M$ the fluid density referred to the maximum value: N_M, which corresponds to the bumper-to-bumper setting of vehicles;

$u\ (t, x) = V(\tau t, \ell x)/V_M$ the dimensionless mass velocity referred to its maximum admissible value: V_M.

If the reference time is selected as follows

$$\frac{\tau V_M}{\ell} = 1 \quad \Rightarrow \quad \tau = \frac{\ell}{V_M}\,, \tag{8.2.4}$$

then Eqs. (8.2.3) become

$$\begin{cases} \dfrac{\partial n}{\partial t} + \dfrac{\partial}{\partial x}(nu) = 0\,, \\[2ex] \dfrac{\partial u}{\partial t} + u\,\dfrac{\partial u}{\partial x} = F\,(n, u, \ldots)\,, \end{cases} \tag{8.2.5}$$

where F is the force density acting on the fluid of vehicles referred to the dimensionless variables.

As above mentioned, particular relations for the mass velocity in the case of scalar models, and for the mass velocity and the force term, in the case of vector models, are to be proposed on the basis of experimental observations. In both cases, they are obtained as mathematical interpretations of the ability and attitude of the driver to adjust the vehicle response to local flow conditions. Vector models may be considered as improvements of the scalar ones because increasing the number of equations increases the number of parameters to be identified.

8.2.1 Scalar hydrodynamic models

Scalar models are described by a single scalar equation, namely the first one of (8.2.5), that rules the cars number evolution with respect to time and space. Considering that this equation involves both n and u, a self-consistent model can be obtained only if a constitutive relation (that is, a phenomenologic relation) can be proposed to link u to n and its derivatives.

The following models, [LEa], [KEa,b], [KUa], [WHa] and [DEa], can be classified as scalar models and be grouped into three main classes:

Model E. *The evolution model for the mass density n is obtained by means of the mass conservation equation together with a phenomenologic relation of the type*

$$u_e = u_e(n)\,, \tag{8.2.6}$$

where u_e is the equilibrium velocity, only depending on the density n, instantaneously reached by the driver. The phenomenologic relation (8.2.6) has to be recovered from identification with experimental data. According to the first of Eq. (8.2.5), the formal structure of the model is

$$\frac{\partial n}{\partial t} + g_e(n)\frac{\partial n}{\partial x} = 0\,, \tag{8.2.7}$$

where

$$g_e(n) = u_e(n) + n\,\frac{\partial u_e}{\partial n}(n)\,. \tag{8.2.8}$$

Model W. *The evolution model for the mass density n is obtained by means of the mass conservation equation wherein not only a phenomenologic relation of the type reported in Eq. (8.2.6) is used, but also a nonlinear diffusion term is added of the type*

$$\varepsilon\frac{\partial}{\partial x}\left(k(n)\frac{\partial n}{\partial x}\right)\,, \qquad 0 < \varepsilon \ll 1\,.$$

A possible form for the function k is $k(n) = n(1-n)$, which gives to the model the following form

$$\frac{\partial n}{\partial t} + g_e(n)\frac{\partial n}{\partial x} = \varepsilon n(1-n)\frac{\partial^2 n}{\partial x^2} + \varepsilon(1-2n)\left(\frac{\partial n}{\partial x}\right)^2\,. \tag{8.2.9}$$

In the case of linear diffusion $k = 1$, and the model is simpler:

$$\frac{\partial n}{\partial t} + g_e(n)\frac{\partial n}{\partial x} = \varepsilon\frac{\partial^2 n}{\partial x^2}\,, \tag{8.2.10}$$

Model D. *The evolution model for the mass density n is again obtained by the diffusionless mass conservation equation together with a phenomenologic relation $u_e = u_e(n^*)$ of the type reported in Eq. (8.2.6). However, in this case, the equilibrium velocity u_e, instantaneously reached by the driver, is assumed to depend on a local fictitious density n^*. The artificial density n^* is meant to take into account the impact on human reactions of the actual traffic conditions, and is considered to be a function not only of the real density, but also of the density gradient. In particular, it is assumed that*

$$u_e = u_e(n^*)\,, \qquad n^* = n\left[1 + \eta(1-n)\frac{\partial n}{\partial x}\right]\,, \tag{8.2.11}$$

where η, with $0 < \eta < 1$, has to be recovered from experimental data. Hence, if the density grows with the space variable then the driver ***feels*** *a local density which is larger than the real one. The closer the real density is to its maximum value, the lower the increase of the fictitious density is with respect to the real one. Then the formal model reads*

$$\frac{\partial n}{\partial t} + u_e\left(n\,,\frac{\partial n}{\partial x}\right)\frac{\partial n}{\partial x} + n\,\frac{\partial}{\partial x}\,u_e\left(n\,,\,\frac{\partial n}{\partial x}\right) = 0\,. \tag{8.2.12}$$

In the literature, [WHa], [KUa], a typical expression of the equilibrium velocity for the E and W models is the following

$$u_e = u_e(n) = \left(1 - n^{1+\alpha}\right)^{1+\beta}\,, \qquad \alpha, \beta \geq 0\,, \tag{8.2.13}$$

that in the relatively simple case (however consistent with experimental data) of $\alpha = \beta = 0$ reduces to

$$u_e = u_e(n) = (1 - n)\ . \tag{8.2.14}$$

Another possible form of Eq. (8.2.7) was proposed by Kerner and Konhäuser [KEa]:

$$u = u_e(n) = h(n) - h(1)\,, \tag{8.2.15}$$

where

$$h(n) = \frac{1}{1 + \exp\left[b_1(n - b_2)\right]}\,, \qquad b_1, b_2 \in \mathbf{R}\,. \tag{8.2.16}$$

We remark that the solutions obtained by models of the class W will have a smoother behavior than those obtained by models of the class E and hence are, possibly, a more realistic description of the physical system to be described.

Additional experiments may improve the analytic simulation of $u_e = u_e(n)$. In particular, they should be developed to identify the parameters related to the personal interpretation of local fictitious density from the viewpoint of the driver.

If the simple expression (8.2.14) is used, the above models can be specialized. The E-model reads

$$\frac{\partial n}{\partial t} + (1-2n)\,\frac{\partial n}{\partial x} = 0\,. \tag{8.2.17}$$

The W-model reads

$$\frac{\partial n}{\partial t} + (1-2n)\frac{\partial n}{\partial x} = \varepsilon n(1-n)\frac{\partial^2 n}{\partial x^2} + \varepsilon(1-2n)\left(\frac{\partial n}{\partial x}\right)^2 , \tag{8.2.18}$$

and the D-model takes the expression

$$\frac{\partial n}{\partial t} + (1-2n)\frac{\partial n}{\partial x} = \eta(n^2-n^3)\frac{\partial^2 n}{\partial x^2} + \eta(2n-3n^2)\left(\frac{\partial n}{\partial x}\right)^2 . \tag{8.2.19}$$

8.2.2 Vector hydrodynamic models

Basic assumption of this class of models is that the velocity u is a variable of the problem at the same level as the mass density, and both are considered as functions of x and t. Their dynamics is described by the two coupled equations (8.2.5).

In the literature (refer again to the reviews [KLa], [KLc]) it is generally assumed that the force density term F is decomposed into two parts: $F = F_1 + F_2$. The first one is a relaxation term towards an equilibrium velocity; the second one is an acceleration term due to the action of the fluid on the driver. In other words, the two terms are related, respectively, to the active and passive behavior of the driver.

The first term is modelled as follows

$$F_1(n,u) = \nu(n)\,\big(u_e(n) - u\big) \tag{8.2.20}$$

where ν is a certain collision frequency and u_e the equilibrium density previously defined. Under the assumption that ν is proportional to n through a relaxation time $T_r = 1/\nu_0$, and making use of the expression for u_e already proposed in the scalar case (see Eq. (8.2.13)), one has

$$F_1(n,u) = \nu_0\big((1-n)-u\big)\,, \tag{8.2.21}$$

where, for simplicity of notations, we used $\alpha = \beta = 0$.

The second term is modelled by assuming that a suitable gradient of the local density induces the driver to accelerate or decelerate. That is

$$F_2(n,u) = -\frac{\theta_0}{n}\frac{\partial n}{\partial x} \tag{8.2.22}$$

where θ_0 is generally taken to be constant, although some authors suggest to replace θ_0 by some function of n, say $\theta = \theta(n)$.

However, as discussed in [KLa], the above modelling gives unrealistic descriptions of the flow conditions under strong changes of densities. Hence a velocity diffusion term is introduced and the following explicit model obtained

$$\begin{cases} \dfrac{\partial n}{\partial t} + \dfrac{\partial}{\partial x}(un) = 0\,, \\[2ex] \dfrac{\partial u}{\partial t} + u\,\dfrac{\partial u}{\partial x} = \nu_0\,\big((1-n)-u\big) - \dfrac{\theta_0}{n}\,\dfrac{\partial n}{\partial x} + \dfrac{\mu}{n}\,\dfrac{\partial^2 u}{\partial x^2}\,. \end{cases} \tag{8.2.23}$$

It is plain that also in this case the artificially introduced term (which is nonlinear if μ is a nontrivial function of n) induces some unphysical energy dissipation.

Alternatively, as proposed in [DEa], the expression for F_1 and F_2 may be generalized by introducing some apparent local density n_* obtained as a function of the mass density, the mass density gradient, and the velocity gradient:

$$n_* = n\left[1 + \eta_1(1-n)\,\frac{\partial n}{\partial x} + \eta_2(1-n)\,\frac{\partial u}{\partial x}\right]\,. \tag{8.2.24}$$

The above structure is developed by linking the hydrodynamic equations to the heuristic modelling of driver's behavior. It is clear that it can be done in several ways and hopefully improved. We feel that further and deeper descriptions of the psyco-physiological behavior of the drivers are appropriate. Some features of this type of modelling are described in Section 3 of [KLa].

On the other hand, a better knowledge of the driver's behavior is also of interest when the alternative point of view is used; namely, the framework that describes the dynamics of each vehicle by use of ordinary differential equations. In this case the driving force is obtained as an output of the interaction model between individuals, the dynamics of the whole system is described by a large number of differential equations, and the macroscopic quantities are recovered by suitable averages over the solutions.

In fact, several arguments may be sustained which limit the validity of the hydrodynamic phenomenologic modelling. In particular, the following are of relevance:

- Very unlikely vehicle flow is continuous. Distances between pairs of vehicles can be large compared with their dimensions.
- Vehicle dynamics is not simply based on purely mechanistic laws. On the contrary, it is substantially influenced by the individual behavior of each driver. The driver program, and its modifications due to traffic conditions, must be included into the hydrodynamic equations.

Therefore we omit going into the details of the above outlined continuous models. In fact, rather than on the macroscopic phenomenological approach, we are interested in the derivation of hydrodynamic equations starting from the kinetic modelling.

8.3 Kinetic Traffic Flow Models

In this section, generalized models of the Boltzmann type will be derived. The main aspects of the problem, that we feel appropriate to be noticed at first, are the following:

- the features that characterize the physical system are typical of the Boltzmann equation;
- interactions among the vehicles are ruled by phenomenologic laws that may be outlined in a way similar to that of classical mechanics;
- human behavior plays a drastically important role in the system evolution.

With a notation similar to that used in the preceding section, let the state of a vehicle along the road be modelled by its position $x \in [0,1]$ and velocity $v \in [0,1]$, i.e., let the vehicle state vector be given by $\mathbf{u} = (x, v) \in [0,1]^2$. We assumed a normalization procedure on the vehicle state variables similar to that followed in the preceding section with reference to the mass variables.

We are concerned with the derivation of an evolution equation for the one vehicle statistical distribution function (with respect to the random variables x, v) parametrized by the scalar variable t

$$f : (t, x, v) \in \ [0,1]^3 \to f(t, x, v) \in \mathbb{R}_+ \,. \tag{8.3.1}$$

When f is integrable one can recover, as usual, the macroscopic observables as moments of the distribution f. In particular, the marginal density

$$\rho(t, x) = \int_0^1 f(t, x, v)\, dv \tag{8.3.2}$$

may be read as the vehicle ***concentration*** or ***number density*** and, hence, $\rho(t, x)dx$ gives the expected number of cars on the road tract dx centered at x at time t. Clearly, $\rho(t, x)$ may be required to satisfy additional conditions such as: $\rho(t,x) \le \rho_M$, where the constant ρ_M denotes the maximum number of cars per unit length, and as such equals the bumper to bumper density N_M and the inverse of the average car length. In deriving the macroscopic equations starting from the kinetic picture the ratio $\rho(t,x)/\rho_M$ has to be identified with the normalized mass density $n = n(t,x)$ of the preceding section.

In the same way, the ***mean velocity*** is given by

$$E[v](t,x) = \frac{1}{\rho(t,x)} \int_0^1 v\, f(t,x,v)\, dv\,, \tag{8.3.3}$$

and this is the velocity that has to be identified with the mass velocity u of the preceding section.

Additional interesting macroscopic quantities in traffic flow theory are the ***speed variance***

$$\mathcal{V}(t,x) := \frac{1}{\rho(t,x)} \int_0^1 [v - E[v](t,x)]^2\, f(t,x,v)\, dv\,, \tag{8.3.4}$$

the ***local flow***

$$\mathcal{Q}(t,x) := \rho(t,x)\; E[v](t,x)\,, \tag{8.3.5}$$

and the so-called ***speed pressure***

$$\mathcal{P}(t,x) := \rho(t,x)\mathcal{V}(t,x) = \int_0^1 [v - E[v](t,x)]^2\, f(t,x,v)\, dv\,. \tag{8.3.6}$$

The analogy with classical quantities of mechanics, such as momentum, energy, and pressure, is evident. It has been sustained however, see [PRb] Section 3.5, that in traffic flow theory the only meaningful conserved quantity should be the density. Hence, the most general equilibrium solution should only depend on the density value. On the other hand, as mentioned above, to derive the hydrodynamic traffic flow description from the kinetic one, the equilibrium hypothesis is required, meaning that the quantities that characterize the system are so slowly varying with respect to space and time that the same description holds as in the time independent homogeneous unforced case, but with the local values of the said quantities instead of the conserved values. In traffic flow theory, therefore, only the first moment might independently be assigned, and any expectations weighted by the local equilibrium distribution function f, necessarily of the form $f = f[\rho(t,x), v]$.

As usual, the development of kinetic modelling and simulation goes through the following subsequent steps:

- Derivation of an evolution equation for the distribution function f;
- Computation of f as a solution of initial and/or boundary value problems;
- Recovering macroscopic quantities, such as the moments of f, as stated in Eqs.(8.3.2)–(8.3.6).

The additional problem of deriving the hydrodynamic equations for the quantities n and u as an asymptotic limit to the continuum description needs to develop an asymptotic theory addressed to the hydrodynamic description.

This section will be subdivided into three parts: the first reports about Prigogine's model, the pioneer kinetic traffic model, and its substantial improvement proposed by Paveri Fontana. The second is a survey of some modifications, and developments, of the previous models. Its contents may be seen as a set of examples, among several others, that show the methodological aspects of kinetic modelling. The third is a review of the mathematical results on the qualitative and quantitative analysis of the evolution equation.

In all cases, the crucial aspect is the description and simulation of the link between vehicle dynamics and dynamics induced by driver's personal behavior. Emphasis will be reserved to this aspect in the chapter.

8.3.1 From Prigogine's to Paveri Fontana's modelling

Modelling traffic flow in a Boltzmann like manner was initiated by Prigogine and Herman. Their merit is to formulate suitable assumptions for the above mentioned interactions; they may be found in the note [PRa] and in various papers collected in the book [PRb]. Afterwards, several authors developed interesting improvements of their model.

A concise description of Prigogine's model is presented here, together with its first substantial revision proposed by Paveri Fontana [PAa]. We first summarize the fundamental assumptions of the original model, then a brief discussion about its merits and conceivable criticisms is given, also in the light of the improvement proposed in [PAa].

In a way similar to the case of hydrodynamic models, the starting point towards modelling is the interaction between mechanics and human behavior. The driver is willing to adjust its velocity, by either increasing or decreasing it, towards a certain desired program. In addition, velocity may change, in fact it may only decrease, due also to the interaction with the heading vehicle. In both cases, the rate of change depends on the density.

Prigogine's model basic assumptions are:

Assumption 8.3.1. *The flow is one-dimensional, and each vehicle is modelled as a point, i.e., the length of each vehicle is negligible with respect to the length of the road, although a maximal density* ρ_M *is considered. The state of each vehicle at time* t *is defined by its position* $x \in [0,1]$ *and velocity* $v \in [0,1]$. *The state of the system is given by the distribution function* $f = f(t,x,v)$ *such that* $f(t,x,v)dx\,dv$ *assigns the number of vehicles that at the time* t *have a state in the phase-space volume* $dx\,dv$ *centered at* (x,v).

Assumption 8.3.2. *The evolution of* f *is ruled by a balance equa-*

tion, generated by vehicles interactions, according to the scheme

$$\frac{\partial f}{\partial t} + v\,\frac{\partial f}{\partial x} = J_P[f]\,. \tag{8.3.7}$$

The term $J_P[f]$ accounts for the rate of change of f due not only to the mechanics of the interactions between vehicles with different velocities, but also to the behavior of the drivers and to their spontaneous speed changes.

The following assumptions model the driver's behavior.

Assumption 8.3.3. *The operator J_P is the sum of two terms*

$$J_P[f] = J_r[f] + J_i[f]\,, \tag{8.3.8}$$

where J_r is the relaxation term, which accounts for the speed change towards a certain ***program*** *of velocities independent of local concentration, and J_i is the term due to the (slowing down) interaction between vehicles.*

The term J_r is related to the fact that each driver (whatever its speed) has a program in terms of a desired velocity, let $f^* = f^*(t, x, v)$ denote the ***desired-velocity distribution function***, meaning that $f^*(t, x, v)dx\,dv$ gives the number of vehicles that, at time t and position in dx at x, have the desire to reach a velocity in dv at v. The driver's desire also consists in reaching this velocity within a certain relaxation time T_r, related to the normalized density and equal for each driver.

Prigogine's relaxation term is defined by

$$J_r[f](t, x, v) = \frac{1}{T_r[f]}\,\left(f^*(t, x, v) - f(t, x, v)\right), \tag{8.3.9}$$

where

$$T_r[f](t, x) = \tau\,\frac{\rho[f](t, x)}{\rho_M - \rho[f](t, x)}\,, \tag{8.3.10}$$

where τ is a constant, and ρ and ρ_M are defined above, see Eq. (8.3.2).

The term J_i is due to the interaction between a trailing (test) vehicle and its heading (field) vehicle. It accounts for the changes in $f(t,x,v)$ caused by a breaking of the test vehicle due to an ***interaction*** with a field vehicle, and it contains a gain term when the test vehicle has velocity $w > v$, and a loss term when the field vehicle has velocity $w < v$. Moreover, J_i is proportional to the probability P that the fast car may pass the slower one, and which is related to the normalized density and equal for each driver.

Prigogine's interaction term is defined by

$$J_i[f](t,x,v) = (1 - P[f])f(t,x,v)\int_0^1 (w-v)f(t,x,w)\,dw\,, \quad (8.3.11)$$

where

$$P[f](t,x) = 1 - \frac{1}{\rho_M}\,\rho[f](t,x)\,. \quad (8.3.12)$$

Assumption 8.3.4. *Molecular chaos is assumed, that is:*

$$f_2(t,x,v,x,w) = f(t,x,v)f(t,x,w)\,.$$

According to the above assumptions, the explicit mathematical model which is then derived consists in Eq. (8.3.7) wherein the interaction term is given by

$$J_P[f](t,x,v) = \frac{1}{\tau}\,\frac{\rho_M - \rho[f](t,x)}{\rho[f](t,x)}\,(f^*(t,x,v) - f(t,x,v))$$
$$+\,\frac{\rho[f](t,x)}{\rho_M}\,f(t,x,v)\int_0^1 (w-v)f(t,x,w)\,dw\,, \quad (8.3.13)$$

and again the density ρ is given by Eq. (8.3.2).

Referring to the classical Boltzmann equation and to generalized Boltzmann models [BLa], the above traffic flow model can be classified as a phenomenologic kinetic model, since it is derived without taking into a detailed account the microscopic interactions between vehicles.

The above aspect will be discussed later. Here it is worth mentioning, as it has been noted in [PAa], that the relaxation term $J_r[f]$ becomes meaningless when the vehicle density n tends to zero, and it is plain that this contradiction needs to be eliminated in order that the model may also be valid in the low density limit. In fact, all technical generalizations developed after the pioneer papers [PRa,b] aim to settle this particular question.

However, this is not the only criticism that can be risen. Indeed, the modelling should also take into account the fact that vehicles may occasionally be involved in high concentration traffic (traffic jams), so that the diluted gas assumption, typical of the Boltzmann equation, has altogether to be put in question. Practical utility of traffic models is closely related to their capability of describing high concentration flows.

The first substantial modification of the model was proposed by Paveri Fontana [PAa]. He criticizes the relaxation term (8.3.9) by showing that it has some unacceptable consequences. Therefore, the desired velocity v^* is assumed to be an independent variable of the problem, and a generalized one vehicle distribution function $g = g(t, x, v; v^*)$ is introduced to describe the distribution of vehicles at (t, x) with speed v and desired speed v^*. Hence the distribution f^* that concerns the desired speed and distribution f that concerns the actual speed are given by

$$f^*(t, x, v^*) = \int_0^1 g(t, x, v; v^*)\, dv\,, \tag{8.3.14}$$

and

$$f(t,x,v) = \int_0^1 g(t,x,v;v^*)\,dv^*\,. \tag{8.3.15}$$

The evolution equation, which now refers to the generalized distribution function g, is again determined by equating the transport term on g to the sum of the slowing down interaction term and relaxation term. That is, one again has

$$\frac{\partial g}{\partial t} + v\frac{\partial g}{\partial x} \;=\; J_{PF}[g] \;=\; J_i[g] + J_r[g]\,. \tag{8.3.16}$$

The interaction term has the same structure as in the previous case; however, now, the operators apply to g, and the passing probability P is assumed to be also a function of a certain critical density ρ_c. Paveri Fontana's interaction term is defined by

$$J_i[g](t,x,v;v^*) = (1-P[f])f(t,x,v)\int_v^1 (w-v)g(t,x,w;v^*)\,dw$$

$$-\,(1-P[f])g(t,x,v;v^*)\int_0^v (v-w)f(t,x,w)\,dw\,, \tag{8.3.17}$$

where

$$P[f](t,x) = (1-\rho[f](t,x)/\rho_c)H(\rho_c - \rho[f](t,x))\,, \tag{8.3.18}$$

and H is the Heaviside function.

The relaxation towards a certain program of velocities is related to vehicles acceleration. This is taken into account by means of a relaxation time T_r that is seen as a function of the passing probability P, and hence of the density. Paveri Fontana's relaxation term is defined by

$$J_r[g](t,x,v;v^*) = -\frac{\partial}{\partial v}\left(\frac{v^*-v}{T_r[f]}\,g(t,x,v;v^*)\right), \tag{8.3.19}$$

where

$$T_r[f](t,x) = \tau \frac{1 - P[f](t,x)}{P[f](t,x)} =: \tau \frac{\rho[f](t,x)}{\rho_c - \rho[f](t,x)} . \qquad (8.3.20)$$

In conclusion, Paveri Fontana's evolution model is given by equations (8.3.15) - (8.3.20).

As in the case of Prigogine's model, the operator J_{PF} takes into account the driver's behavior in a phenomenologic way. Moreover, concerning the collision operators, both the models are based upon heuristic arguments. Therefore, they are not fully justified at a microscopic level, and allow several different improvements.

For instance, a modification of Prigogine's model was proposed by Lampis [LAa,b] in order to take into account the effect of queuing among vehicles. The analysis refers to stationary homogeneous flow, and allows some comparisons with experimental data. The main idea consists in adding the model with a distribution function $g = g(t, x, v)$ for (the leader of vehicles in) a queue, and introducing an interaction term of the form (8.3.14) to account for the interactions between vehicles with distribution g and vehicles with distribution f.

In all cases, the models here cited have the same structure of the Boltzmann equation, however the similarity does not go beyond the formal aspects of the evolution equation, and a detailed microscopic modelling has not been given.

8.3.2 Developments in kinetic modelling

Extending the two pioneering models we have just recalled, recently some authors developed kinetic models based upon a detailed microscopic description of the pair interactions. The resulting evolution equations are suitable statistical balances, and under this aspect they closely resemble Boltzmann equation. Indeed, the collision operator is the difference between a gain and a loss term defined on the basis of the microscopic interactions.

Here we briefly outlines some of the efforts recently made, see e.g., [NEa], [KLa], [KLb], to provide an abstract yet rigorous approach to the derivation of kinetic equations. We present them as examples of possible developments of Prigogine's and Paveri Fontana's theory, and no purpose of completeness is claimed.

The collision operator is modelled, on a mechanical pairwise interaction ground, by analyzing each driver short-range reactions to neighborhood vehicles rather than interpreting his overall behavior. The interactions are strictly pairwise in that the test vehicle only reacts to what happens in its immediate headings. This makes the model more similar to the Boltzmann kinetic scattering equation, and in some sense closer to the follow-the-leader point of view. Under these aspects, this class of models may be considered as the real starting points for further developments.

Models stemming from the microscopic description of pair interactions have the advantages that comparisons with experimental data (and organization of suitable experiments) can be arranged being only related to the mentioned microscopic behaviors. Moreover, stationary solutions, that are of great relevance in the analysis of traffic flow, are direct predictions of the model.

This approach is in opposition with phenomenologic modelling which, on the contrary, can be validated only by experiments that evolve on the time scale of the distribution function dynamics, and are quite difficult to be organized.

The new approach was introduced and developed by various authors, e.g., Nelson [NEa], Klar and Wegner [KLa,b], and [WEa], who were able to exploit the advantages of a modelling based upon describing the short range interactions.

However, modelling microscopic interactions is certainly not a simple task. It requires detailed analysis of vehicle dynamics and driver's reactions, together with the organization of specifically related experiments.

The problem is to find suitable expressions for the post-interaction velocities v' and w' which, in the microscopic modelling, are directly related to the pre-interaction ones: v and w. Furthermore, if high densities must be taken into account, then modifications of the interaction frequency should be included in a way similar to that followed in deriving Enskog equation.

In [NEa] Nelson proposes a model, the author himself calls it ***a traffic flow caricature***, which suffers from some severely restrictive assumptions on the dynamics allowed to each driver. Some of them are: zero passing probability; a unique value for the desired velocity which coincides with the highest possible speed v_M; a unique minimal headway distance; a particular kind of vehicular chaos; instantaneous changes of speed; instantaneous reactions to any circumstances. These last two assumptions implying two different time scales: an immediate time scale and an evolution time scale. Correspondingly, the author obtains a bimodal equilibrium distribution, centered at the desired (maximal) speed $v_M = 1$ and at the zero speed, which is somewhat surprising. In fact this bimodal distribution is related to the time scale of immediate reactions.

On the other hand, he is the first author who treats the speeding-up event in essentially the same way as the slowing-down event, in that he proposes a ***table of the truth*** of outgoing velocities in response to each possible incoming circumstance. In other words, a transition probability density $\psi(v, v')$ is *a priori* sketched. Hence, he doesn't need to introduce any relaxation term such as those of the Prigogine's and Paveri Fontana's models.

In more details, Nelson model is based on several assumptions described in what follows.

The set of possible values for the test vehicle outgoing velocity after a change-in-speed event, or ***interaction***, is restricted to $\{0, v, v_H, v_M = 1\}$, where v_H denotes the heading vehicle velocity.

Fundamental to any interaction process, because it triggers the occurrence of the change-in-speed event, is the ***minimal headway*** $\xi(v)$. The headway distance ξ is assumed to be a strictly increasing function of just one velocity, mostly the heading vehicle speed. A suitable ***headway probability density*** $p(h|t, x, v)$ is introduced to account for the probability that the headway distance may be less than a certain h. The transition probability ψ is then constructed depending upon the various possible values of h with respect to the desired value: $\xi(v)$.

In addition, the following technical assumptions are proposed:

$$p(h|t,x,v) = p(h|t,x)\,, \tag{8.3.21}$$

$$p(h|t,x) = \begin{cases} 1 - exp(-\rho(t,x)(h-\xi(0))) & \text{if } h > \xi(0), \\ 0 & \text{if } h \le \xi(0), \end{cases} \tag{8.3.22}$$

$$g(h, v_H; t, x, v) = \frac{f(t,x,v_H)}{\rho(t,x)} \frac{dp}{dh}(h|t,x)\,, \tag{8.3.23}$$

where the vehicle density $\rho(t,x)$ is defined as in Eq. (8.3.2), and where $g(h, v_H; t, x, v)$ denotes the probability density that a trailing vehicle has a leading vehicle at headway h with velocity v_H.

The explicit model proposed by Nelson is

$$\frac{\partial f}{\partial t} + v\,\frac{\partial f}{\partial x} = J_N[f] := \delta(t) J_s[f] \; + \; J_l[f] \tag{8.3.24}$$

where the delta function factor $\delta(t)$ accounts for the short times response and where, omitting the (t,x) dependence, the long times term is given by

$$J_l[f](v) = f(v)\left(\int_0^1 (p(\xi(v')) - p(\xi(v)))|v' - v| f(v')\,dv'\right)\,. \tag{8.3.25}$$

If $q_1 := (1 - p(\xi(1)))$ denotes the minimal probability of non passing, the short times term is

$$J_s[f](v) = \frac{1}{\rho}\left[-f(v)\left(\int_0^v p(\xi(v'))f(v')\,dv' + \rho q_1\right)\right.$$

$$+\,\delta(v-1)q_1\left(\int_0^{1^-} \rho\, f(v_2)\,dv_2 + H(v-1)f(1)q_1\right)$$

$$\left. +\,\delta(v)\left(\int_{0+}^1 f(v_1 -)\int_0^{v_1} p(\xi(v_2))f(v_2)\,dv_2\,dv_1\right)\right]. \tag{8.3.26}$$

Starting from Nelson's model, to release some of its restrictive assumptions and to gain a more general and flexible structure, Klar and Wegener adopt in [WEa] essentially the same mathematical structure of gain and loss term that is familiar in the Boltzmann theory.

They introduce an ***outgoing velocity probability density function*** $\sigma(v; v_1, v_2)$, which is related to the event that vehicle "1" (instantaneously) changes his speed from v_1 to v because of an interaction with its heading vehicle "2" at velocity v_2. In fact, they consider several possible ***headway thresholds*** $h_i(v_1, v_2)$, $i = 1, \ldots, r$, and hence several possible functions σ_i corresponding to various different kinds of interactions.

Interactions are assumed to happen only when the headway distance ***crosses*** any of the thresholds values h_i's. In this way fast time scales may be avoided, probabilities $q_i(v_2, \tau; t, x_1, v_1)$ may be introduced concerning the event that vehicle 1 interacts with vehicle 2 in the time interval $[t, t+\tau]$ due to the headway crossing the value $h_i(v_1, v_2)$. The time rate of the interaction probability density may then be written as

$$\sum_{i=1}^{r} \frac{dq_i}{d\tau}(v_2, 0; t, x_1, v_1)\,. \tag{8.3.27}$$

This leads to the following detailed model

$$\frac{\partial f}{\partial t} + v\frac{\partial f}{\partial x} = J_{KW}[f] = G[f] - L[f]\,, \tag{8.3.28}$$

where

$$G[f](t, x, v) = \sum_{i=1}^{r} \int_{(v_1, v_2)\in\Omega_i} f(t, x, v_1)\sigma_i(v; v_1, v_2) \times \frac{dq_i}{d\tau}(v_2, 0; t, x, v_1)\, dv_1\, dv_2\,, \tag{8.3.29}$$

$$L[f](t,x,v) = f(t,x,v)\sum_{i=1}^{r}\int_{(v,v_2)\in\Omega_i}\frac{dq_i}{d\tau}(v_2,0;t,x,v)\,dv_2\,, \quad (8.3.30)$$

and Ω_i denotes the set of all the velocity pairs that allow i-th interaction.

On the other hand, the probability densities q_i's, $i = 1,\ldots,r$, may be specialized as follows

$$q_i(v_2,\tau;t,x,v_1) = \chi_{\Omega_i}(v_1,v_2)\,sign(v_1 - v_2) \times \int_{h_i}^{h_i+(v_1-v_2)\tau} g(h,v_2,t,x,v_1)\,dh \quad (8.3.31)$$

where $h_i = h_i(v_1,v_2)$, χ_{Ω_i} is the characteristic function of the set Ω_i, and the function g denotes the number density of the leading vehicles that are at headway h, and velocity v_2, from a trailing vehicle at (t,x,v); i.e.:

$$f_2(t,x,v_1,x+h,v_2) = g(h,v_2;t,x,v_1)f(t,x,v_1)\,,$$

where f_2 denotes the pair distribution function.

On the function $g = g(h,v_2;t,x,v_1)$ same special assumptions are then stated, similar to those already used in Nelson model, and necessary to obtain a closed equation for f. They lead to the following detailed form

$$g(h_i(v_1,v_2),v_2;t,x,v_1) = f(t,x,v_2)\;k(h_i(v_1,v_2);\rho(t,x)) \equiv f(t,x,v_2)\;\exp(-\rho(t,x)(h_i(v_1,v_2) - \xi(0)))\,, \quad (8.3.32)$$

where $\xi(0)$ denotes the speed zero headway distance.

The explicit model of Klar and Wegener is finally obtained by Eq. (8.3.28) wherein the gain and loss terms are specialized into the following ones, again omitting the (t,x) dependence,

$$G[f](v) = \sum_{i=1}^{r}\int_{(v_1,v_2)\in\Omega_i} f(v_1)\sigma_i(v;v_1,v_2)|v_1 - v_2|$$

$$\times \; g(h_i(v_1, v_2), v_2; v_1)\, dv_1\, dv_2 \,, \tag{8.3.33}$$

$$L[f](v) = f(v) \sum_{i=1}^{r} \int_{(v,v_2)\in\Omega_i} |v - v_2| g(h_i(v, v_2), v_2; v)\, dv_2, \tag{8.3.34}$$

and where the form (8.3.32) is used for the function g.

Numerical simulations show that, even in the easy case of one slowing down and one speeding up thresholds h_i's, qualitative behavior of the homogeneous equilibrium distribution obtained by this model is in good agreement with that observed in real experiments.

Subsequently, and following the same lines, Klar and Wegener in [KLb] introduce an Enskog-like approach and re-propose the interaction terms (8.3.33),(8.3.34) as follows

$$G[f](v) = \sum_{i=1}^{r} \int_{(v_1,v_2)\in\Omega_i} |v_1 - v_2| \sigma_i(v; v_1, v_2)$$
$$\times \; f_2(t, x, v_1, x + h_i(v_1, v_2), v_2)\, dv_1\, dv_2 \,, \tag{8.3.35}$$

and

$$L[f](v) = \sum_{i=1}^{r} \int_{(v,v_2)\in\Omega_i} |v - v_2| f_2(t, x, v, x + h_i(v, v_2), v_2)\, dv_2 \,, \tag{8.3.36}$$

where $f_2(t, x_1, v_1, x_2, v_2)$ denotes the two-vehicle distribution function, a test vehicle at (x_1, v_1) and a heading vehicle at (x_2, v_2). Again one has: $f_2(t, x, v_1, x+h, v_2) = g(h, v_2; t, x, v_1) f(t, x, v_1)$; however, now, the function g is such that

$$g(h, v_2; t, x, v_1) = f(t, x + h, v_2)\; k(h, \rho(t, x))\,.$$

They consequently deduce a Navier-Stokes-like equation of the form

$$\begin{cases} \dfrac{\partial n}{\partial t} + \dfrac{\partial}{\partial x}(nu) = 0\,, \\[2ex] \dfrac{\partial}{\partial t}(nu) + \dfrac{\partial}{\partial x}(p_e(n) + nu^2) + a_e(n)\dfrac{\partial n}{\partial x} = \dfrac{nu_e(n) - nu}{T_e(n)}\,, \end{cases} \tag{8.3.37}$$

where the anticipation coefficient a_e, and the inverse T_e of the interaction frequency, are conveniently defined, and where all the quantities $u_e(n)$, $\mathcal{P}_e(n)$, $a_e(n)$ and $T_e(n)$ are determined from the equilibrium solution of the homogeneous kinetic equations. Equations (8.3.37) have the same qualitative terms as those already used in the literature. Also in this case, the model is validated by numerical simulations that seem to closely represent the backward effect which is proper of high density traffic jam conditions.

All the above models have been derived by assuming that all vehicles are free to move without external actions of any kind, either human or mechanical. The distribution function is modified only by pair interactions.

This picture may considerably be improved if one assumes that the flow conditions induce an action equivalent to an external force. In this case, the formal evolution equation reads

$$\frac{\partial f}{\partial t} + v\frac{\partial f}{\partial x} + \frac{\partial}{\partial v}(fF[f]) = J[f]\,, \tag{8.3.38}$$

where $F[f]$ is the external force and $J[f]$ the collision operator to be defined according to one of the above models. We remark that the forcing term F can be modelled according to the phenomenologic analysis described in Section 8.2 and, in particular, fictitious quantities may play a role to take into account the different human reactions to different traffic conditions.

8.3.3 Evolution problems

The qualitative analysis of the initial value problem for kinetic traffic flow models is an interesting and challenging problem. It was initiated, on the basis of the semigroup theory, by Belleni Morante both with reference to Prigogine's model [BEa–BEc], and to Paveri Fontana's model [BAa], [BEd].

In [BEb] the solution to the initial-boundary value problem of Prigogine's model is sought in the form

$$f(t,x,v) = \varphi(v) + \psi(t,x,v)\,, \tag{8.3.39}$$

where φ is the solution of the steady state problem obtained by equating to zero Prigogine collision operator $J_P[f]$, and where ψ is a perturbation term, though not necessarily small. In more details, the Authors analyze the following problem for $f = f(t,x,v)$ on $[0,\infty) \times [0,1] \times [v_m, 1]$, where $v_m \geq 0$,

$$\begin{cases} \dfrac{\partial f}{\partial t} + v\dfrac{\partial f}{\partial x} = J_P[f] \\ f(t,x,v) = \varphi(v) + \psi(t,x,v) \end{cases} \tag{8.3.40}$$

with the auxiliary conditions

$$\begin{cases} \lim\limits_{t\to 0^+} \psi(t,x,v) = \psi_0(x,v)\,, \qquad \text{on} \quad [0,1]\times[v_m,1]\,, \\ \\ \lim\limits_{x\to 0^+} \psi(t,x,v) = 0\,, \qquad \text{for} \quad t \geq 0, \quad v \in [v_m, 1]\,. \end{cases} \tag{8.3.41}$$

The analysis consists in the research of mappings from $[0,\infty)$ to the Banach space $\mathbf{X}$ of real functions $h : [0,1]\times[v_m,1] \to \mathbb{R}_+$ such that $h \in C\,([0,1]\times[v_m,1])$ and that $h(0,\cdot) \equiv 0$. The space $\mathbf{X}$ is equipped with the sup norm. Moreover, it is assumed that a nonnegative function $\kappa = \kappa(v)$ exists such that

$$f^*(t,x,v) \;=:\; \kappa(v)\,\rho(t,x) \tag{8.3.42}$$

where f^* denotes, as above, the desired-velocity distribution function, and ρ the density.

On $\mathbf{X}$, the following operators are introduced

$$K[h](x,v) := \frac{\kappa(v)}{T} \int_{v_m}^{1} h(x,v')\, dv'; \tag{8.3.43}$$

$$H[h](x,v) := \int_{v_m}^{1} (v'-v)h(x,v')\, dv'; \tag{8.3.44}$$

$$B[h](x,v) := -v\, \frac{\partial h}{\partial x} - \frac{1}{T}\, h\,; \tag{8.3.45}$$

where

$$\mathcal{D}(B) := \{h \in \mathbf{X} \,\big|\, B[h] \in \mathbf{X}\} \tag{8.3.46}$$

is the domain of the operator B, and T is a given relaxation time.

On use of the latter, the problem is rephrased as an abstract version of a nonlinear initial value problem on the perturbation ψ:

$$\frac{\partial \psi}{\partial t} = B[\psi] + K[\psi] + (Q - Q_0)\, \varphi\, H[\varphi] \\ + Q\left(\psi H[\varphi] + \varphi H[\psi] + \psi H[\psi]\right), \tag{8.3.47}$$

where $Q := (1-P)$, and P is the passing probability, possibly time dependent. Equation (8.3.47) is endowed with the following auxiliary conditions

$$\psi_0,\, \psi(t) \in \mathcal{D}(B), \quad \text{and} \quad \lim_{t\to 0+} \|\psi(t) - \psi_0\| = 0\,. \tag{8.3.48}$$

On this problem the following properties are proved to be true:

i) B is the generator of a strongly continuous translation semigroup $Z := \{Z_t\}_{t\geq 0}$ on $\mathbf{X}$:

$$Z_t[h] = \exp\left(-\frac{t}{T}\right)\, h(s, x - vt, v)$$

if $x - vt \geq 0$, and

$$Z_t[h] = 0$$

if $x - vt < 0$;

ii) both K and H are bounded on $\mathbf{X}$ and map the domain $\mathcal{D}(B)$ into itself.

Then, as in [KAa], the following necessary and sufficient condition are recognized for the strongly continuous solutions of Eqs. (8.3.47), (8.3.48), omitting the x, v-dependence,

$$\begin{aligned}\psi(t) =& Z_t[\psi_0] + (Q - Q_0) \int_0^t Z_{(t-s)} \; [\varphi H[\varphi]] \; ds \\ &+ \int_0^t Z_{(t-s)} \Big[K[\psi] + Q(\psi H[\varphi] \\ &+ \varphi H[\psi] + \psi H[\psi]) \Big] (s) \, ds \end{aligned} \tag{8.3.49}$$

wherein the integrals are strong Riemann integrals.

Then, with $\psi^{(0)}(t) := Z_t[\psi_0]$ as starting point, Eq. (8.3.49) is used to define a procedure of successive approximations towards the function $\psi(t)$. Owing to the properties above for the operators H, K, and B, the procedure is convergent on $\mathbf{X}$ provided that t belongs to some interval $[0, t_0]$ for a certain finite $t_0 \in \mathbb{R}$. (The smaller the product $Q\psi_0|v_1 - v_2|^2$, the greater t_0 may be chosen.)

Regularity properties of the solution $\psi = \psi(t; Q, \psi_0)$ are also proved with respect to the parameter Q, such as its strong differentiability. Hence, the meaningfulness follows of a Taylor-like expansion of ψ in powers of Q.

Finally it is seen that if the initial perturbation ψ_0 is suitably small then the vehicle distribution decays, as $t \to \infty$, towards the equilibrium distribution $\varphi(v)$.

The analysis has also been developed for Paveri Fontana's model with the same objectives on existence and stability of solutions. The same methods sketched above are followed, although suitable developments are necessary to deal with the greater difficulties that this second model involves. Hence, there is no point in providing here further details on these results. We simply mention that the analysis in [BEa], [BEb] is valid for a further large variety of models as well, including those proposed in [KLa], [KLb], and [WEa].

On the other hand, several interesting problems remain open, and can be regarded as a challenging task for applied mathematicians.. For instance, it is certainly of interest to refer the above qualitative analysis to the computational schemes developed towards simulation. This implies, for instance, investigating on the regularity properties of the solutions, with respect to the x-variable, on the basis of the regularity assumptions on initial data.

8.4 Perspectives

As we have seen in the preceding sections, modelling traffic flow is an interesting research field related to applied sciences. Thus, further research activity is justified to improve the state of the art. In principle, research perspectives should refer both to continuous and kinetic modelling. On the other hand, we shall concentrate on kinetic models not only because this topic is consistent with the aims of this book, but also because the authors' conviction is that continuous modelling should be obtained, with the methods of asymptotic analysis, as a limit framework when the distances between vehicles become small with respect to the road length (hydrodynamic limit). Some of the remarks developed in what follows are also induced by very recent research activity in the field [KLc].

With this in mind, the topics that we feel fundamental in view of research perspectives are the following:

- Analysis of the modelling and possible experiments;
- Statement of the mathematical problems, and development of their analysis towards optimization and control;

- Asymptotic analysis towards continuum description.

The last two arguments are self explanatory, and we do not add many further comments. Both of them are intimately connected with the particular kind of mathematical modelling that has been developed, and follow methods and procedures that may considerably vary from a model to another.

Referring instead to the ***modelling***, further analysis may be developed starting from the various papers by Nelson, Klar and Wegener we discussed in Section 8.3. Fundamental point is modelling all the details of vehicles interactions, such as encounter rates, transition probabilities, response to traffic conditions, passing dynamics. Indeed, these quantities are crucially affected by the natural driver behavior, who subjectively modifies any purely mechanistic behavior. On the contrary, all the above mentioned authors have exploited ideas quite close to kinetic theory of gases.

Developments in modelling being addressed towards the human reaction description, the experiments should be organized to validate the microscopic modelling. Kinetic equations are indeed derived with technical calculations once the basic assumptions on the microscopic behavior are stated.

Afterwards, as it is described in [KLc] for the general case of multilane modelling, simulation begins with constructing in details all the quantities that are necessary to write, and compute, each of the terms that appears in the kinetic equation. Then, the procedure must be repeated for the macroscopic equation. Guiding line is to compute them starting from a coherent microscopic dynamics, based on the individual cars behaviors, rather than using phenomenologic *a priori* relations.

Concerning kinetic traffic flow equations, the interaction term is, in all cases, a balance between a gain term and a loss term of vehicles from a certain velocity (or position and velocity) state. Therefore, the quantities that must be simulated ultimately are: on the one hand some particular vehicles distribution functions, such as the leading-vehicles distribution or the distribution of vehicles with a certain desired speed or simply the vehicle distribution function itself; on the other hand some special coefficients or probability densities, such as the passing, breaking, changing lane, and accelerating probabilities. The said functions are distributions with respect to the vehicle velocities, and generally at equilibrium conditions.

All the various quantities to be determined possibly depend not only on the position and velocity of the test vehicle, but also on other conditioning events, such as the particular values and kinds of the velocity distribution function, or of the desired velocity distribution function. Therefore some *a priori* stochastic assumptions are also necessary, together with some driving rules derived from the usual traffic laws and based on the standard driver behavior.

It is clear, in addition, that two possible scenarios need to be separately developed: the space homogeneous case and the space dependent one. While the first one is of interest for stationary solutions in unlimited or periodic traffic flow conditions, the second one may be developed to take into account effects such as bottlenecks or traffic jams. Yet, in this case, one has to release the assumption of cars of zero length.

Once the said quantities have been recovered and computed, the kinetic interaction operator may be completely simulated and the kinetic equation solved by means of some convenient numerical approximation method. Analysis can be developed either by the computational scheme proposed in [KLb], which is a generalization of discretization methods applied to the Boltzmann equation, or by collocation interpolation method. Actually, the evolution problem is not so hard as it is in the case of the Boltzmann equation. The reason is that we are dealing with problems in one space dimension, and with bounded velocities. Therefore, we expect that several efficient computational schemes can be developed.

It is plain that necessary conditions for the simulation to proceed are:

- that the above first-step computations, at the microscopic level, be validated by qualitative and quantitative comparison with the experimental observations that are available in the literature;
- that the results of the second-step computations, at the kinetic level, be themselves compared with well known and accepted data.

In particular, the kinetic equation is at first used to compute the stationary homogeneous distribution function at various values for the total car density. Then it is compared with the microscopic results. Finally it is used to construct the fundamental diagram.

At last, and in particular in the spatially inhomogeneous case, a third step may be performed by solving the fluid dynamics equations that are derived for the continuous model. The various coefficients that appear in these equations must be computed, as mentioned above, under the assumption that the macroscopic quantities coincide with statistical averages. Hence, they are obtained by making use of the solution of the second step of the simulation. For instance, traffic flow density, flux, and traffic pressure are the first three moments of the distribution function f. The continuous model solutions are, ultimately, the only quantities that may be compared with acceptable experimental measurements.

8.5 References

[BAa] Barone E. and Belleni Morante A., A nonlinear initial value problem arising from kinetic theory of vehicular traffic, *Transp. Theory Stat. Phys.*, **7**, (1978) 61–79.

[BEa] Belleni-Morante A., Stability in kinetic theory of vehicular traffic, *Meccanica*, **1**, (1973) 11–15.

[BEb] Belleni-Morante A. and Barone E., Nonlinear kinetic theory of vehicular traffic, *J. Math. Anal. Appl.*, **47**, (1974) 443–457.

[BEc] BELLENI-MORANTE A. and PAGLIARINI A., Two-group kinetic theory of vehicular traffic, *Meccanica*, **3**, (1974) 151–156.

[BEd] BELLENI-MORANTE A. and FROSALI G., Global solution of a nonlinear initial value problem of vehicular traffic, *Boll. U.M.I.*, **5**, (1977) 71–81.

[BEf] BELLENI MORANTE A., **Applied Semigroup and Evolution Equations**, Oxford Univ. Press, Oxford, (1980).

[BEe] BELLENI-MORANTE A. and MCBRIDE A., **Applied Nonlinear Semigroup**, Wiley, New York, (1998).

[BLa] BELLOMO N., LE TALLEC P., and PERTHAME B., On the Solution of the Nonlinear Boltzmann Equation, *ASME Review*, **48** (1995) 777–794.

[CMa] CHORIN A. and MARSDEN J., **Mathematical Introduction to Fluid Dynamics**, Springer, Heidelberg, (1979).

[DEa] DE ANGELIS E., Hydrodynamic flow models, *Mathl. Comp. Modelling*, **29**, (1999), 83–96.

[GAa] GAZIS D.C., HERMAN R. and ROTHERY R., Nonlinear follow-the-leader model for traffic flow, *Opn. Res.*, **9**, (1961) 545–574.

[KAa] KATO T., Nonlinear evolution equations in Banach spaces, in **Proceedings of AMS Symposium in Applied Mathematics**, AMS, **17**, (1965) 50–67.

[KEa] KERNER B. and KONHÄUSER P, Cluster effect in initially homogeneous traffic flow, *Physical Review E*, **48**, (1993) 2335–2338.

[KEb] KERNER B. and KONHÄUSER P, Structure and parameters of clusters in traffic flow, *Physical Review E*, **50**, (1994) 54–83.

[KLa] KLAR A., KÜNE R. D., and WEGENER R., Mathematical models for vehicular traffic, *Surveys Math. Ind.*, **6**, (1996) 215–239.

[KLb] KLAR A. and WEGENER R., Enskog-like kinetic models for vehicular traffic, *J. Stat. Phys.*, **87**(1/2), (1997) 91–114.

[KLc] KLAR A. and WEGENER R., Kinetic traffic flow models, in **Modeling in Applied Sciences: A Kinetic Theory Approach**, Bellomo N. and Pulvirenti M. Eds., (1999).

[KUa] KÜNE R. D. and RÖDIGER M.B., Macroscopic simulation model for freeway traffic with jams and stop-start waves, in **Proceedings of the 1991 Winter Simulation Conference**, Nelson B.L., Kelton W.D., and Clark G.M. Eds., (1991) 762–771.

[LAa] LAMPIS M., On the kinetic theory of traffic flow in the case of a non negligible number of queuing vehicles, *Transp. Science*, **12** (1978) 16–28.

[LAb] LAMPIS M., On the Prigogine theory of traffic flow: Driver's program independent of concentration, *Meccanica*, (1977), 187–193.

[LEa] LEUTZBACK W., **Introduction to the Theory of Traffic Flow**, Springer, Heidelberg, (1988).

[NAa] NAGATANI T., Spreading of traffic jam in a traffic flow model, *J. Phys. Soc. Japan*, **62** (1993), 1085–1088.

[NEa] NELSON P., A kinetic model of vehicular traffic and its associated bimodal equilibrium solution, *Transp. Theory Stat. Phys.*, **24** (1995), 383–409.

[NSa] NAGEL K. and SCHRECKENBERG M., A cellular automaton model for freeway traffic, *J. Phys. I France*, **2** (1992), 2221–2229.

[PAa] PAVERI FONTANA S.L., On Boltzmann like treatments for traffic flow, *Transp. Res.*, **9** (1975), 225–235.

[PRa] PRIGOGINE I., RESIBOIS P., HERMAN R., and ANDERSON R., On a generalized Boltzmann like approach for traffic flow, *Acad. Royale de Belgique*, **48**, (1962) 805–814.

[PRb] PRIGOGINE I. and HERMAN R., **Kinetic theory of vehicular traffic**, Elsevier, New York, (1971).

[WEa] WEGENER R. and KLAR A., A kinetic model for vehicular traffic derived from a stochastic microscopic model, *Transp. Theory Stat. Phys.*, **25**(7), (1996) 785–798.

[WHa] WHITHAM G., **Linear and Nonlinear Waves**, J. Wiley, London (1978).

Chapter 9

Dissipative Kinetic Models for Disparate Mixtures

9.1 Introduction

In general, one of the interesting developments of the Boltzmann equation towards technological applications is the analysis of flows of disparate gas mixtures, say bubbles, droplets, clusters or solid particles. Derivation of kinetic models is possible if the physical conditions characterizing the system are similar to those of the Boltzmann equation. In particular, the dimensions of interacting particles should be small compared with their mean free path, which should be of the same order as the characteristic length of the obstacles to the flow and/or of the vessel containing the system, i.e., ***rarefied gas*** conditions.

This chapter deals with a generalization of the Boltzmann equation referred to gas mixtures of particles undergoing dissipative collisions. The colliding objects may either be of particles, or molecular clusters. The collision scheme is such that interactions preserve mass and momentum, while energy is dissipated. If an interaction occurs between clusters, then the collision can eventually modify their size.

The class of models proposed in this chapter is closely related to the original Boltzmann model. Indeed, the evolution equation refers to the distribution in the phase space (position and velocity) of a system consisting of a large number of particles. Microscopic interactions are governed by the equations of classical mechanics.

Referring to general aspects, we focus our attention on the three features that characterize the system and, ultimately, the mathematical model. They are: ***cluster size***, ***dissipative interactions***, and ***mixtures***.

The first feature is the interacting particles ***size***. One can model each element either as a point, and obtain a Boltzmann-like model, or as an element with finite size, and then interactions occur at a certain distance from the centers of mass of the particles. In this case the model is developed in the same fashion as the Enskog equation. This also means that a pair of correlation functions should be introduced to modify the collision frequency due to the fact that particles occupy part of the volume available for the collisions.

The second feature refers to particles ***interactions***, which may be perfectly elastic or, in alternative, energy dissipative. In the latter case mass and momentum are still preserved, while energy is dissipated. Indeed, it is reasonable to include dissipation for particles which have a large dimension with respect to molecules. For instance interactions of clusters may be such that part of the kinetic energy is converted into vibrational energy with some dissipation.

Finally the last feature refers to ***mixture*** modelling. The model can either refer to a system of a finite number of species each constituted by particles of the same size or to a disparate mass mixture of particles with random distributed sizes. Additional interesting physical features can be taken into account. For instance, one may include condensation vaporization phenomena and so on. Here, we provide the main lines which lead to the modelling of a disparate mixture of particles with dissipation of energy.

A preliminary analysis of mixtures with dissipative collisions was developed in [BLb] dealing with the derivation of kinetic equations and hydrodynamics. Indeed, it was shown that a new hydrodynamics can be derived with a dissipative term in the energy equation. The reference analysis was developed in the papers [ESa] and [ESb] dealing with evolution equations with non elastic collisions and spin. Kinetic models for mixtures undergoing dissipative collisions with fragmentation coagulation phenomena have been proposed by Slemrod and Qi [SLa], and Slemrod [SLb], based on the discrete Boltzmann equation, a model of the kinetic theory of gases [GAa] where the particles are allowed to attain only a finite number of velocities. Kinetic models for clusters undergoing dissipative collisions are derived in [LOa]. Further, it seems, see [AAa] and [AAb], that this type of models can provide useful information on the mass distribution of Saturn's ring. Other fields of interest can be recovered in the modelling of chemically reacting gases [GIa], [GRa].

The above indications motivate the contents of this chapter devoted to the development of a kinetic theory for mixture of gases with energy dissipative collisions that preserve mass and momentum.

The contents is organized in six sections.

Section 9.1 is this introduction.

Section 9.2 deals with the analysis of the collision mechanics for a disparate mixture of clusters that undergo energy dissipative collisions and may modify their size due to fragmentation and condensation phenomena.

Section 9.3 deals with the derivation of kinetic equations for mixtures of clusters.

Section 9.4 deals with the derivation of kinetic equations for a disparate mixtures of particles with continuous mass distribution.

Section 9.5 provides an overview on some analytic results.

Section 9.6 contains a discussion on the applications and research perspectives.

9.2 Dissipative Collision Dynamics

This section deals with an analysis of the mechanics of elastic and inelastic collisions in a mixture of clusters with mass distribution within a certain range of admissible masses.

Consider a mixture of spherical clusters constituted by ν, with $\nu = 1, \ldots, n$, elementary particles with mass m. Rotational dynamics is neglected. In other words, it is assumed that the size ν of each cluster is small, so that it can be dealt with as material particles.

In particular, we consider the collision of a ***test*** particle of ***mass*** $\nu_v m$ with a ***field*** particle of mass $\nu_w m$. As usual, pre-collision ***velocities*** will be denoted $\mathbf{v}$ and $\mathbf{w}$, while $\mathbf{v}'$ and $\mathbf{w}'$ are the post-collision velocities of the test and field particles respectively. Similarly, the primes indicate the post-collision sizes ν'_v and ν'_w of the clusters respectively corresponding to the pre-collision sizes ν_v and ν_w.

It is assumed, here in after, that all collisions preserve mass and momentum, while energy may be dissipated. Moreover, it is convenient to distinguish among:

a) ***Totally conservative collisions*** which preserve both energy and the cluster sizes: $\nu'_v = \nu_v$ and $\nu'_w = \nu_w$.

b) ***Cluster conservative*** and ***energy dissipative collisions*** which preserve the sizes of the interacting clusters, $\nu'_v = \nu_v$ and $\nu'_w = \nu_w$, while the energy is dissipated.

c) ***Cluster destructive*** and ***energy dissipative collisions*** which preserve neither energy nor the interacting clusters sizes $\nu'_v \neq \nu_v$ and $\nu'_w \neq \nu_w$. Yet, the overall size is not modified by the collision: $\nu'_v + \nu'_w = \nu_v + \nu_w$.

9.2.1 Cluster conservative collisions

Consider first ***cluster and energy conservative collisions***. In this case, the following conservation equations can be stated

$$\begin{cases} \nu_v = \nu'_v , \qquad \nu_w = \nu'_w \\ \nu_v \mathbf{v} + \nu_w \mathbf{w} = \nu_v \mathbf{v}' + \nu_w \mathbf{w}' \\ \nu_v |\mathbf{v}|^2 + \nu_w |\mathbf{w}|^2 = \nu_v |\mathbf{v}'|^2 + \nu_w |\mathbf{w}'|^2 . \end{cases} \tag{9.2.1}$$

The analysis of ***totally conservative collisions***, such that neither momentum nor energy are modified, is classical in the literature.

As known, the post-collision velocities $\mathbf{v}', \mathbf{w}'$ are obtained in terms of a two-dimensional parameter which identifies the scattering directions. The relations are similar to those of the classic Boltzmann equation, see Chapter 7, with some technical modifications to take into account interactions of different masses:

$$\begin{cases} \mathbf{v}' = \mathbf{v} + \dfrac{2\nu_w}{\nu_v + \nu_w} \langle \mathbf{w} - \mathbf{v}\,,\, \mathbf{n} \rangle \, \mathbf{n}\,, \\ \\ \mathbf{w}' = \mathbf{w} - \dfrac{2\nu_v}{\nu_v + \nu_w} \langle \mathbf{w} - \mathbf{v}\,,\, \mathbf{n} \rangle \, \mathbf{n}\,. \end{cases} \tag{9.2.2}$$

Consider now ***cluster conservative and energy dissipative collisions***. Following [ESa], [ESb], let $\beta \in [0, \frac{1}{2})$ be the parameter which characterizes energy dissipation. The post-collision velocities are given by

$$\begin{cases} \mathbf{v}' = \mathbf{v} + \dfrac{2\nu_w}{\nu_v + \nu_w} (1 - \beta) \langle \mathbf{w} - \mathbf{v}\,,\, \mathbf{n} \rangle \, \mathbf{n}\,, \\ \\ \mathbf{w}' = \mathbf{w} - \dfrac{2\nu_v}{\nu_v + \nu_w} (1 - \beta) \langle \mathbf{w} - \mathbf{v}\,,\, \mathbf{n} \rangle \, \mathbf{n}\,. \end{cases} \tag{9.2.3}$$

System (9.2.3), as shown in [ESb], admits an inverse solution which gives the velocities $\mathbf{v}_*$ and $\mathbf{w}_*$ which are necessary, as pre-collision velocities, to produce $\mathbf{v}, \mathbf{w}$ as output velocities in a dissipative collision with given $\mathbf{n}$, β, ν_v, and ν_w:

$$\begin{cases} \mathbf{v}_* = \mathbf{v} + \dfrac{2\nu_w}{\nu_v + \nu_w} \left(\dfrac{1 - \beta}{1 - 2\beta} \right) \langle \mathbf{w} - \mathbf{v}\,,\, \mathbf{n} \rangle \, \mathbf{n}\,, \\ \\ \mathbf{w}_* = \mathbf{w} - \dfrac{2\nu_v}{\nu_v + \nu_w} \left(\dfrac{1 - \beta}{1 - 2\beta} \right) \langle \mathbf{w} - \mathbf{v}\,,\, \mathbf{n} \rangle \, \mathbf{n}\,. \end{cases} \tag{9.2.4}$$

In particular, $\beta = 0$ corresponds to completely elastic collisions. For $\beta = 0$ the standard notations $\mathbf{v}'$ and $\mathbf{w}'$ will be used: one obtains

$$\beta = 0 \quad \Rightarrow \quad \mathbf{v}' = \mathbf{v}_*\,, \quad \mathbf{w}' = \mathbf{w}_*\,.$$

Note that for every $\beta \in [0, \frac{1}{2})$, mass and momentum are preserved in the collisions. On the other hand, energy is preserved only for $\beta = 0$. Indeed, one has

$$\nu_v|\mathbf{v}|^2 + \nu_w|\mathbf{w}|^2 = \nu_v|\mathbf{v}_*|^2 + \nu_w|\mathbf{w}_*|^2 + 4\beta(1-\beta)\frac{\nu_v\nu_w}{\nu_v + \nu_w}|\langle\mathbf{n}\,,\,\mathbf{w}-\mathbf{v}\rangle|^2\,. \tag{9.2.5}$$

9.2.2 Cluster destructive collisions

Consider now the case of ***cluster destructive and energy dissipative collisions***; they are such that the cluster sizes change, during the collisions phenomena, due to coagulation and fragmentation procedure. Yet, the total mass is still preserved. In this case one has to assume suitable coagulation–fragmentation relations $\nu'_v = \nu'_v(\nu_v, \nu_w)$, and $\nu'_w = \nu'_w(\nu_v, \nu_w)$ such that

$$\nu_v + \nu_w = \nu'_v(\nu_v, \nu_w) + \nu'_w(\nu_v, \nu_w)\,. \tag{9.2.6}$$

The modelling can be developed according to the following:

Assumption 9.2.1. *There exists a **critical size** ν_c, with $1 < \nu_c < n$, such that if*

$$(\nu_v - \nu_c)(\nu_w - \nu_c) \geq 0\,,$$

*the collision is **cluster conservative**, and if*

$$(\nu_v - \nu_c)(\nu_w - \nu_c) < 0\,,$$

*the collision is **cluster destructive**. In this last case, **coagulation** and **fragmentation** phenomena happen so that*

$$\begin{cases} \nu_v < \nu_c < \nu_w & \Rightarrow \quad \nu'_v = \nu_v + b\,, \quad \nu'_w = \nu_w - b\,, \\ \nu_w < \nu_c < \nu_v & \Rightarrow \quad \nu'_v = \nu_v - b\,, \quad \nu'_w = \nu_w + b\,, \end{cases} \tag{9.2.7}$$

where

$$1 \leq b = \min\{|\nu_v - \nu_c|\ |\nu_w - \nu_c|\}\ .$$

Assumption 9.2.2. *The* ***dissipation coefficient*** $\beta = \beta(\nu_v, \nu_w) \in [0, 1/2)$ *satisfies the condition*

$$\langle \mathbf{n}\ ,\ (\mathbf{v}' - \mathbf{w}')\rangle = (1 - 2\beta)\left\langle \mathbf{n}\ ,\ \left(\frac{\nu_w}{\nu'_w}\mathbf{w} - \frac{\nu_v}{\nu'_v}\mathbf{v}\right)\right\rangle . \tag{9.2.8}$$

The first assumption can be justified on a phenomenologic basis according to a modelling of the phenomenon such that the cluster with a size smaller than ν_c has a trend to increase its size when colliding with a cluster with a size larger than ν_c. A simplified description, which is commonly used, [SLa], [SLb], takes $b = 1$

$$\begin{cases} \nu_v < \nu_c < \nu_w & \Rightarrow \quad \nu'_v = \nu_v + 1\,, \quad \nu'_w = \nu_w - 1\,, \\ \nu_w < \nu_c < \nu_v & \Rightarrow \quad \nu'_v = \nu_v - 1\,, \quad \nu'_w = \nu_w + 1\,. \end{cases} \tag{9.2.9}$$

The second assumption can be regarded as a generalization of the one proposed in [ESa], [ESb] and re-visited in this section. Consider the following auxiliary collision

$$(\nu'_v \mathbf{v}_0\ ,\ \nu'_w \mathbf{w}_0) \to (\nu'_v \mathbf{v}'\ ,\ \nu'_w \mathbf{w}')\,, \tag{9.2.10}$$

which is such that the pre-collision moments coincide with those of the true collision

$$\nu_v \mathbf{v} = \nu'_v \mathbf{v}_0\,, \qquad \nu_w \mathbf{w}_w = \nu'_w \mathbf{w}_0\,,$$

and that the overall momentum is also preserved. Denote with

$$\mathcal{E}_0 = \frac{1}{2}(\nu'_v \mathbf{v}_0^2 + \nu'_w \mathbf{w}_0^2)\,, \tag{9.2.11}$$

and with

$$\mathcal{E}' = \frac{1}{2}(\nu'_v \mathbf{v}'^2 + \nu'_w \mathbf{w}'^2)\,, \tag{9.2.12}$$

respectively, the pre-collision and post-collision energies. It follows from the auxiliary collision that

$$\mathcal{E}_0 - \mathcal{E}' = 2\beta(1-\beta)\frac{\nu'_v \nu'_w}{\nu_v + \nu_w}\left(\langle \mathbf{n}\,,\, \mathbf{w}_0 - \mathbf{v}_0\rangle\right)^2\,, \tag{9.2.13}$$

where $\beta \in [0, \frac{1}{2})$. Therefore β is identified as the energy dissipative coefficient of the auxiliary collision (9.2.10).

Technical calculations yield the expressions of the post-collision velocities

$$\begin{cases} \mathbf{v}' = \dfrac{\nu_v}{\nu'_v}\mathbf{v} + \dfrac{2\nu'_w}{\nu_v + \nu_w}(1-\beta)\left\langle \left(\dfrac{\nu_w}{\nu'_w}\mathbf{w} - \dfrac{\nu_v}{\nu'_v}\mathbf{v}\right)\,,\, \mathbf{n}\right\rangle \mathbf{n}\,, \\[2em] \mathbf{w}' = \dfrac{\nu_w}{\nu'_w}\mathbf{w} - \dfrac{2\nu'_v}{\nu_v + \nu_w}(1-\beta)\left\langle \left(\dfrac{\nu_w}{\nu'_w}\mathbf{w} - \dfrac{\nu_v}{\nu'_v}\mathbf{v}\right)\,,\, \mathbf{n}\right\rangle \mathbf{n}\,. \end{cases} \tag{9.2.14}$$

The derivation of kinetic equations requires further analysis concerning the calculation of the Jacobian of the velocity transformation and the identification of the ordered pairs of post-collision cluster sizes that generate the pre-collision sizes. In particular, referring to the map $(\mathbf{v}\,,\mathbf{w}) \to (\mathbf{v}'\,,\mathbf{w}')$, it can be shown that the Jacobian is defined by

$$\frac{d(\mathbf{v}'\,,\mathbf{w}')}{d(\mathbf{v}\,,\mathbf{w})} = (1-2\beta)\left(\frac{\nu_v \nu_w}{\nu'_v \nu'_w}\right)^3\,. \tag{9.2.15}$$

To complete the analysis of this type of collisions, consider the inverse problem

$$(\nu_{v_*}\mathbf{v}_*\,, \nu_{w_*}\mathbf{w}_*) \to (\nu_v \mathbf{v}\,, \nu_w \mathbf{w})\,, \tag{9.2.16}$$

where the particles pre-collision sizes ν_{v_*}, ν_{w_*} and velocities $\mathbf{v}_*, \mathbf{w}_*$, which generate as an output the post-collision sizes ν_v, ν_w and the velocities $\mathbf{v}, \mathbf{w}$, are to be computed.

The analysis is in two steps, the first one refers to the sizes and the second one to the velocities.

Step 1. The problem consists in computing the set $\mathcal{D}(\nu_v, \nu_w)$ of those pairs (ν_{v_*}, ν_{w*}) which generate a given pair (ν_v, ν_w). In general, according to Eq. (9.2.6)

$$\begin{aligned}
&\mathcal{D}(\nu_v, \nu_w) \subseteq \mathcal{B}(\nu_v, \nu_w) \\
&= \left\{ (\nu_{v_*}, \nu_{w*}) \in \{1, \ldots, n\}^2 \;\middle|\; \nu_{v_*} + \nu_{w*} = \nu_v + \nu_w \right\}. \qquad (9.2.17)
\end{aligned}$$

The detailed computation of the term $\mathcal{D}$ can be developed following Assumption 9.2.1

$$\begin{cases}
\nu_v = \nu_w = \nu_c \Rightarrow \mathcal{D}(\nu_v, \nu_w) = \mathcal{B}(\nu_v, \nu_w), \\
\nu_v = \nu_c < \nu_w \quad \text{or} \quad \nu_v < \nu_w = \nu_c \\
\qquad \Rightarrow \mathcal{D}(\nu_v, \nu_w) = \left\{ (\nu_{v_*}, \nu_{w^*}) \in \mathcal{B}(\nu_v, \nu_w) \;\middle|\; \nu_{v_*} < \nu_{w*} \right\}, \\
\nu_c = \nu_w < \nu_v \quad \text{or} \quad \nu_w < \nu_v = \nu_c \\
\qquad \Rightarrow \mathcal{D}(\nu_v, \nu_w) = \left\{ (\nu_{v_*}, \nu_{w*}) \in \mathcal{B}(\nu_v, \nu_w) \;\middle|\; \nu_{w*} < \nu_{w*} \right\}, \\
(\nu_v - \nu_c)(\nu_w - \nu_c) > 0 \Rightarrow \mathcal{D}(\nu_v, \nu_w) = (\nu_v, \nu_w), \\
(\nu_v - \nu_c)(\nu_w - \nu_c) < 0 \Rightarrow \mathcal{D}(\nu_v, \nu_w) = \emptyset.
\end{cases} \qquad (9.2.18)$$

Step 2. The problem consists in determining the pre-collision velocities $\mathbf{v}_*$ and $\mathbf{w}_*$ related to collision (9.2.16). Calculations similar to those we have seen before yield

$$\begin{cases} \mathbf{v}_* = \left(\dfrac{\nu_v}{\nu_{v_*}}\right)\mathbf{v} + \dfrac{\nu_{w_*}}{\nu_v + \nu_w}\,\dfrac{2(1-\beta)}{1-2\beta}\left\langle \dfrac{\nu_w}{\nu_{w_*}}\mathbf{w} - \dfrac{\nu_v}{\nu_{v_*}}\mathbf{v}\,,\,\mathbf{n}\right\rangle \mathbf{n}\,, \\[2ex] \mathbf{w}_* = \left(\dfrac{\nu_w}{\nu_{w_*}}\right)\mathbf{v} - \dfrac{\nu_{v_*}}{\nu_v + \nu_w}\,\dfrac{2(1-\beta)}{1-2\beta}\left\langle \dfrac{\nu_w}{\nu_{w_*}}\mathbf{w} - \dfrac{\nu_v}{\nu_{v_*}}\mathbf{v}\,,\,\mathbf{n}\right\rangle \mathbf{n}\,. \end{cases} \tag{9.2.19}$$

In addition, one has

$$\langle \mathbf{w}_* - \mathbf{v}_*\,,\,\mathbf{n}\rangle = \frac{1}{1-2\beta}\left\langle \left(\frac{\nu_w}{\nu_{w_*}}\right)\mathbf{w} - \frac{\nu_w}{\nu_{w_*}}\mathbf{v}\,,\,\mathbf{n}\right\rangle . \tag{9.2.20}$$

Finally consider the map $(\mathbf{v}\,,\mathbf{w}) \to (\mathbf{v}_*\,,\mathbf{w}_*)$, the Jacobian is

$$\frac{d(\mathbf{v}_*\,,\mathbf{w}_*)}{d(\mathbf{v}\,,\mathbf{w})} = \frac{1}{(1-2\beta)}\left(\frac{\nu_v\nu_w}{\nu_{v_*}\nu_{w_*}}\right)^3 . \tag{9.2.21}$$

9.3 Kinetic Equations for Mixtures of Clusters

We can now deal with the problem of deriving kinetic equations for the mixture in both cases: cluster conservative and destructive collisions. Consider first the case of ***cluster conservative collisions***, where the collision mechanics is defined by Eqs. (9.2.1)–(9.2.5), and consider the distribution function $f_{\nu_v} = f_{\nu_v}(t, \mathbf{x}, \mathbf{v})$ of the particles of size ν_v. The generalized Boltzmann equation for the mixture is a system evolution equation for the distribution functions f_{ν_v}. One has to take into account collisions of clusters with mass defined by the parameter ν_v and then consider all admissible encounters with the clusters characterized by ν_w. The derivation is a technical development of the one proposed in [LOa].

Kinetic equations will be derived for the hard spheres model. In this case, technical calculations yield

$$\left(\frac{\partial}{\partial t} + \langle \mathbf{v}\,,\,\nabla_{\mathbf{x}}\rangle\right) f_{\nu_v} = \sum_{\nu_w=1}^{n} J_{\nu_v\nu_w}[\mathbf{f}]\,, \tag{9.3.1}$$

where $\mathbf{f} = (f_{\nu_v}, f_{\nu_w})$ and where the operator $J_{\nu_v\nu_w}$:

$$J_{\nu_v\nu_w}[\mathbf{f}] = G_{\nu_v\nu_w}[\mathbf{f}] - L_{\nu_v\nu_w}[\mathbf{f}] \tag{9.3.2}$$

is defined by

$$G_{\nu_v\nu_w}[\mathbf{f}](t,\mathbf{x},\mathbf{v}) = \frac{a^2}{(2\beta-1)^2}\int_{\mathbb{R}^3\times\mathbf{S}^2_+} \langle \mathbf{n}, \mathbf{w}-\mathbf{v}\rangle$$

$$\times\; f_{\nu_v}(t,\mathbf{x},\mathbf{v}_*) f_{\nu_w}(t,\mathbf{x},\mathbf{w}_*)\, d\mathbf{n}\, d\mathbf{w}\,, \tag{9.3.3}$$

$$L_{\nu_v\nu_w}[\mathbf{f}](t,\mathbf{x},\mathbf{v}) = a^2 f_{\nu_v}(t,\mathbf{x},\mathbf{v}) \int_{\mathbb{R}^3\times\mathbf{S}^2_+} \langle \mathbf{n}, \mathbf{w}-\mathbf{v}\rangle$$

$$\times\; f_{\nu_w}(t,\mathbf{x},\mathbf{w})\, d\mathbf{n}\, d\mathbf{w}\,, \tag{9.3.4}$$

where $\mathbf{v}_*, \mathbf{w}_*$ are given by Eqs. (9.2.4), a is the radius of the action domain of the particles, namely the radius of the effective cross section, and $\mathbf{S}^2_+$ is the integration domain of the variable $\mathbf{n}$:

$$\mathbf{S}^2_+ = \left\{\mathbf{n}\in\mathbf{S}^2 \;\middle|\; \langle \mathbf{n}\,,\,\mathbf{w}-\mathbf{v}\rangle \geq 0\right\}. \tag{9.3.5}$$

The derivation of the above equation is based upon the classical phenomenologic Boltzmann assumptions we have already seen in Chapter 7:

a) The loss term is defined by test particles that enter into the action domain of the field particle and hence lose their state;

b) The gain term is computed by all clusters which, due to the collisions, gain the state $\mathbf{v}$;

c) Collisions are instantaneous and local in space, moreover the joint probability in the states $\mathbf{v}$ and $\mathbf{w}$ is factorized;

d) The parameter β does not depend on the size of the interacting pairs.

It is immediate to show that mass and momentum are preserved in the collision, that is

$$\sum_{\nu_v=1}^{n}\sum_{\nu_w=1}^{n}\int_{R^3}(G_{\nu_v\nu_w}[\mathbf{f}] - L_{\nu_v\nu_w}[\mathbf{f}])\,d\mathbf{v} \;=\; 0\,, \tag{9.3.6}$$

and

$$\sum_{\nu_v=1}^{n}\sum_{\nu_w=1}^{n}\int_{R^3}(G_{\nu_v\nu_w}[\mathbf{f}] - L_{\nu_v\nu_w}[\mathbf{f}])\,\nu_v\,\mathbf{v}\,d\mathbf{v} \;=\; 0\,. \tag{9.3.7}$$

On the other hand, energy is not preserved unless $\beta = 0$. In other words, energy is preserved if and only if the collision operator is equal to zero; that is, for all $\nu_v \in I\!N$,

$$\sum_{\nu_w=1}^{n}\int_{R^3}(G_{\nu_v\nu_w}[\mathbf{f}] - L_{\nu_v\nu_w}[\mathbf{f}])\,\nu_v\,|\mathbf{v}|^2\,d\mathbf{v} \;=\; 0\,. \tag{9.3.8}$$

Consider now the case of ***cluster destructive collisions***, where now the collision mechanics is defined by Eqs. (9.2.6)–(9.2.21). Again the Boltzmann equation for a mixture is a system evolution equation for the distribution functions f_{ν_v}. The kinetic equation is characterized by the same structure of Eq. (9.3.1). However, one has to consider the fact that, now, interactions can also modify the clusters' size. In particular, the loss term is computed in the same way as above since it is again defined by test particles that, entering into the action domain of the field particle, lose their state. On the other hand, computation of the gain term requires considering all admissible fragmentation and coagulation of clusters that may generate clusters of size ν_v in the state $\mathbf{v}$. Moreover, one has to take into account that the Jacobian of the map $(\mathbf{v}, \mathbf{w}) \to (\mathbf{v}_*, \mathbf{w}_*)$, see Eq. (9.2.21), depends on the sizes of the colliding clusters.

The evolution equation in the case of hard spheres interaction writes

$$\left(\frac{\partial}{\partial t} + \langle \mathbf{v}, \nabla_{\mathbf{x}} \rangle + \langle \mathbf{F}_{\nu_v}, \nabla_{\mathbf{v}} \rangle \right) f_{\nu_v} = \sum_{\nu_w=1}^{n} J^d_{\nu_v \nu_w}[\mathbf{f}]$$

$$= \sum_{\nu_w=1}^{n} G^d_{\nu_v \nu_w}[\mathbf{f}] - L^d_{\nu_v \nu_w}[\mathbf{f}], \tag{9.3.9}$$

where $\mathbf{F}_{\nu_v}$ is the force acting on the cluster of size ν_v, and where the operator $J^d_{\nu_v \nu_w}$:

$$J^d_{\nu_v \nu_w}[\mathbf{f}] = G^d_{\nu_v \nu_w}[\mathbf{f}] - L^d_{\nu_v \nu_w}[\mathbf{f}] \tag{9.3.10}$$

is defined by

$$G^d_{\nu_v \nu_w}[\mathbf{f}](t, \mathbf{x}, \mathbf{v}) = \frac{a^2}{1 - 2\beta} \sum_{\nu_{v_*}, \nu_{w_*} \in \mathcal{D}(\nu_v, \nu_{v_*})} \int_{\mathbb{R}^3 \times S^2_+} \langle \mathbf{n}, \mathbf{v}_* - \mathbf{w}_* \rangle$$

$$\times \left(\frac{\nu_v \nu_w}{\nu_{v_*} \nu_{w*}}\right)^3 f_{\nu_{v_*}}(t, \mathbf{x}, \mathbf{v}_*)\, f_{\nu_{w_*}}(t, \mathbf{x}, \mathbf{w}_*)\, d\mathbf{n}\, d\mathbf{w}, \tag{9.3.11}$$

$$L^d_{\nu_v \nu_w}[\mathbf{f}](t, \mathbf{x}, \mathbf{v}) = a^2 f_{\nu_v}(t, \mathbf{x}, \mathbf{v})$$

$$\times \int_{\mathbb{R}^3 \times S^2_+} \langle \mathbf{n}, \mathbf{w} - \mathbf{v} \rangle f_{\nu_w}(t, \mathbf{x}, \mathbf{w})\, d\mathbf{n}\, d\mathbf{w}. \tag{9.3.12}$$

In the above equations, $\mathbf{v}_*, \mathbf{w}_*$ are given by Eqs. (9.2.19), while the admissible sets are defined by Eqs. (9.2.18). Also in this case one can show that mass and momentum are preserved in the collision, while energy is preserved only if $\beta = 0$.

The above models have been written for hard-spheres interaction potential; simple technical modifications lead to models with general interaction potential.

9.4 Mixtures with Continuous Mass Distribution

An analysis similar to that we have seen in the preceding section can be developed for a mixture of particles characterized by a continuous mass distribution with values in $[m_m, m_M] \subset \mathbb{R}$. In this case, we consider the relatively simpler problem of interactions, however dissipative, which do not modify the size.

The size of the interacting particles can be identified by the dimensionless parameters

$$\alpha_v = \frac{m_v - m_m}{m_M - m_m}\,, \qquad \alpha_w = \frac{m_w - m_m}{m_M - m_m}\,, \tag{9.4.1}$$

respectively corresponding to test and field particles. The normalization is such that $\alpha_v\,,\,\alpha_w \in [0,1]$.

By means of technical calculations similar to those of Section 9.2, one can show that post-collision velocities are given by

$$\begin{cases} \mathbf{v}' = \mathbf{v} + \dfrac{2\mu + 2\alpha_w}{2\mu + \alpha_v + \alpha_w}(1-\beta)\langle \mathbf{w} - \mathbf{v}\,,\,\mathbf{n}\rangle\,\mathbf{n}\,, \\ \\ \mathbf{w}' = \mathbf{w} - \dfrac{2\mu + 2\alpha_v}{2\mu + \alpha_v + \alpha_w}(1-\beta)\langle \mathbf{w} - \mathbf{v}\,,\,\mathbf{n}\rangle\,\mathbf{n}\,, \end{cases} \tag{9.4.2}$$

where

$$\mu = \frac{m_m}{m_M - m_m}\,.$$

Moreover the velocities $\mathbf{v}_*$ and $\mathbf{w}_*$, which produce the velocities $\mathbf{v}, \mathbf{w}$ in the dissipative collisions are given by

$$\begin{cases} \mathbf{v}_* = \mathbf{v} + \dfrac{2\mu + 2\alpha_w}{2\mu + \alpha_v + \alpha_w} \left(\dfrac{1-\beta}{1-2\beta} \right) \langle \mathbf{w} - \mathbf{v} \,,\, \mathbf{n} \rangle \, \mathbf{n} \,, \\ \\ \mathbf{w}_* = \mathbf{w} - \dfrac{2\mu + 2\alpha_v}{2\mu + \alpha_v + \alpha_w} \left(\dfrac{1-\beta}{1-2\beta} \right) \langle \mathbf{w} - \mathbf{v} \,,\, \mathbf{n} \rangle \, \mathbf{n} \,. \end{cases} \tag{9.4.3}$$

If a continuous mass distribution is assumed, then one deals with a continuous family of distribution functions $\{f_{\alpha_v}(t, \mathbf{x}, \mathbf{v})\}_{\alpha_v \in [0,1]}$. Calculations similar to those we have just seen leads to an evolution equation that takes into account collisions of particles with a mass in the parameter range $d\alpha_v$ at α. All admissible encounters are then considered by integrating over α_w in the domain $[0, 1]$. The model explicitly reads

$$\left(\frac{\partial}{\partial t} + \langle \mathbf{v}, \nabla_{\mathbf{x}} \rangle + \langle \mathbf{F}_{\alpha_v}, \nabla_{\mathbf{v}} \rangle \right) f_{\alpha_v} = \int_0^1 \left(G_{\alpha_v \alpha_w}[\mathbf{f}] - L_{\alpha_v \alpha_w}[\mathbf{f}] \right) d\alpha_w \,, \tag{9.4.4}$$

where $\mathbf{F}_{\alpha_v}$ is the force field acting on the particles of size α_v,

$$G_{\alpha_v \alpha_w}[\mathbf{f}](t, \mathbf{x}, \mathbf{v}) = \frac{a^2}{(2\beta - 1)^2} \int_0^a \int_{\mathbf{R}^3 \times \mathbf{S}_+^2} \langle \mathbf{n}, \mathbf{w} - \mathbf{v} \rangle \times f_{\alpha_v}(t, \mathbf{x}, \mathbf{v}_*) \, f_{\alpha_w}(t, \mathbf{x}, \mathbf{w}_*) \, d\mathbf{n} \, d\mathbf{w} \,, \tag{9.4.5}$$

and

$$L_{\alpha_v \alpha_w}[\mathbf{f}](t, \mathbf{x}, \mathbf{v}) = a^2 f_{\alpha_v}(t, \mathbf{x}, \mathbf{v}) \times \int_0^a \int_{\mathbf{R}^3 \times \mathbf{S}_+^2} \langle \mathbf{n} \,, \mathbf{w} - \mathbf{v} \rangle f_{\alpha_w}(t, \mathbf{x}, \mathbf{w}) \, d\mathbf{n} \, d\mathbf{w} \,. \tag{9.4.6}$$

As already mentioned in the preceding section, Enskog-type models can be derived by introducing suitable pair correlation functions and collisions which take into account the finite dimension of the particles.

9.5 Mathematical Problems

Mathematical problems for kinetic mixtures does not substantially differ from those related to the Boltzmann equation. One has to set properly initial and boundary conditions for each gas component and then analyze the mathematical problems on the basis of the methods reviewed in Chapter 7. If the model is obtained in the framework of the elastic Boltzmann equation, then analysis leads to the same results, reviewed in [BLa], available for the Boltzmann or Enskog equations. Additional analysis has to be developed for models with dissipative collisions.

We refer only to cluster conservative models. The analysis of the Cauchy problem for dissipative models in the nonlinear case is limited to the paper by Esteban and Perthame, where it is shown that the proof by DiPerna and Lions can be generalized to the Boltzmann equation with dissipative collisions and spin. On the other hand, no results are yet available concerning existence of equilibrium solutions and perturbation of equilibrium.

The solution of the above topic would certainly contribute to solving several interesting problems including the rigorous derivation of the hydrodynamic limit which is dealt with, at a formal level, in [BLb]. Specifically, the analysis developed in [BLb] shows that the formal limit yields hydrodynamic equations with energy dissipation. Therefore analytic results on this problem may contribute to a deeper understanding of the foundation of dissipative hydrodynamics.

The analysis of initial-boundary value problems may again be developed with techniques similar to those developed for the Boltzmann equation. However, also in this case, a specific analysis is not really available in the literature. The analysis requires, preliminarily, to state the boundary conditions which should, coherently to the structure of the model, take into account dissipation of the collisions at the wall. This topic has to be regarded as an interesting research field, where a negligible part of the existing literature can be used.

9.6 Perspectives in Modelling

We consider now some perspectives in the modelling of disparate mixtures that undergo elastic and dissipative collisions. The following two topics are proposed, among several others, to the attention of the reader:

• Derivation of the models proposed in Sections 9.2 – 9.4 is based on classical techniques generally used for the phenomenologic derivation of the Boltzmann equation. An alternative approach may start from Liouville equation and develop a hierarchy of equations that model a dynamics of particles that makes no use of the above mentioned simplifications. This program is developed in two papers by Russo and Smereka [RUa], [RUb], which can be regarded as a valuable reference point for further research activity in this field.

Due to this reason, the sequential steps of their analysis is summarized here:

a) A kinetic formulation of a rarefied bubbly flow is developed for a fluid of identical bubbles schematized as rigid spheres. The analysis is developed neglecting bubble collisions as well as the effect of viscosity and gravity.

b) Modelling the interaction of the bubbles with the fluid gives rise to a Hamiltonian system for a constant number of interacting particles. Various interesting interaction models are considered.

c) The Hamiltonian system yields a Vlasov-type equation for the distribution function which is derived from the Liouville equation associated to the Hamiltonian flow.

d) A kinetic type evolution equation, called by the authors: ***Boltzmann- Vlasov equation*** is derived in order to take into account the collisions between bubbles.

e) Fluid dynamic equations are then derived by moments of the Boltzmann Vlasov equation.

The interesting aspect of this approach is its generality. Indeed, it can be considered as a basis for further developments towards different physical systems, say systems of particles with different radii, clusters with vaporization and condensation phenomena, and so on. Recent research activity in this sense has been developed by Jabin and Perthame [JAa], [JAb].

• Additional work needs to be done to deal with models that describe interactions of droplets or bubbles involved in condensation vaporization phenomena. In this case, the size of the interacting particles is not only modified by collisions with other particles, but also evolves according to phase transition phenomena on their surface. Modelling should take into account the above phenomena by suitable phase transition models. This leads to modify both terms on the left and right of the kinetic evolution equation. A contribution to this type of modelling may be developed looking for the links between the models described in this chapter and those dealt with in Chapter 3.

9.7 References

[AAa] ARAKI S. and TREMAINE S., The dynamics of dense particles disks, *Icarus*, **65**, (1986), 83–109.

[AAb] ARAKI S., The dynamics of dense particle disks - Effects of Spin degrees of freedom, *Icarus*, **76** (1988), 182–198.

[BLa] BELLOMO N, in **Lecture Notes on the Mathematical Theory of the Boltzmann Equation**, Bellomo N. Ed., *Advances in Mathematics for Applied Sciences n.25*, World Scientific, London, Singapore, (1995).

[BLb] BELLOMO N., ESTEBAN M., and LACHOWICZ M., Nonlinear kinetic equations with dissipative collisions, *Appl. Math. Letters*, **8** (1995), 47–52.

[ESa] ESTEBAN M. and PERTHAME B., Solutions globale de l'equation d'Enskog modifiée avec collisions elastique ou inelastique, *Comp. Rend. Acad. Sci. Paris*, **309** (1989), 897–902.

[ESb] ESTEBAN M. and PERTHAME B., On the modified Enskog equation for elastic and inelastic collisions. Models with Spin., *Ann. Inst. Poincare*, **8** (1991), 289–308.

[GIa] GIOVANGIGLI V. and MASSOT M., Asymptotic stability of equilibrium states for multicomponent reactive flows, *Math. Models Meth. Appl. Sci.*, **8**, (1998), 251–298.

[GRa] GRÜNFELD C., Non-linear kinetic models with chemical reactions, in **Modelling in Applied Sciences, a Kinetic Theory Approach**, Bellomo N. and Pulvirenti M. Eds., Birkhäuser, Basel, (1999).

[JAa] JABIN P. and PERTHAME B., Notes on mathematical problems on the dynamics of dispersed particles interacting through a fluid, in **Modelling in Applied Sciences: A Kinetic Theory Approach**, Bellomo N. and Pulvirenti M. Eds., Birkhäuser, Basel, (1999).

[LOa] LONGO E. and BELLOMO N., Nonlinear kinetic equations with dissipative collisions, *Appl. Math. Letters*, **12**, (1999), 71–76.

[RUa] RUSSO G. and SMEREKA P., Kinetic theory of bubbly flow I: collisionless case, *SIAM J. Appl. Math.*, **56** (1996), 327–357.

[RUb] RUSSO G. and SMEREKA P., Kinetic theory of bubbly flow II: collisionless case, *SIAM J. Appl. Math.*, **56** (1996), 357–371.

[SLa] SLEMROD M. and QI A., A discrete velocity coagulation fragmentation model, *Math. Models Meth. Appl. Sci.*, **5** (1995), 619–640.

[SLa] SLEMROD M., Metastable fluid flow described via a discrete velocity coagulation fragmentation model, *J. Stat. Phys.*, **83**

(1996), 1067–1108.

Chapter 10

Research Perspectives

10.1 Introduction

The contents of the preceding chapters provided a unified presentation of the methodology, and applications, of generalized kinetic models developed to analyze a large variety of systems of several interacting elements. The aim was to organize and present the mathematical framework, the methodological aspects of modelling, the applications and developments, of a new class of models in applied sciences, based on generalizing ideas that Boltzmann developed in the framework of classical nonequilibrium kinetic theory.

Each of the chapters was related to a specific class of models, and their contents organized along a unified line of presentation:

a) Modelling methods;

b) Mathematical statement of problems, followed by their qualitative analysis;

c) A review of the model applications, followed by an analysis of its conceivable developments.

For all these new models, Boltzmann equation is a fundamental reference point. It is possible, though, to regard it as only a particular, however relevant, element of a broader class of models which refer to social sciences, biology, immunology, physics. This more general framework has been developed along this book, starting from the anticipation given in the review paper [BLb].

Although an effort was made towards completeness, several problems have not been considered. Rather than hiding them, the aim of this chapter is to bring unsolved problems to the attention of interested readers, in a critical framework that may stimulate research activity in this field, and hopefully new developments. For instance, kinetic models with quantum interactions, see e.g., [KEa] and [MAa], are not dealt with in this book. The above topic has a relevance that is undeniable, also in view of its impact with technological applications e.g., in the field of semiconductor devices. On the other hand, it deserves a much larger space than that taken by each of the preceding chapters.

Other topics of the same relevance are cited in the review papers [BLc]; in it new ideas, fields of applications, and interesting research perspectives are gathered showing how the interest for generalized kinetic models is increasing.

This final chapter concentrates on some modelling aspects which have not been discussed in the preceding chapters. Specifically, we consider: on the one hand the possibility of designing discrete models, that have the advantage of being relatively simpler than the continuous ones, on the other hand the possibility of developing a general structure, that may be suitable to include all the models of kinetic type.

The Chapter is divided into four section.

Section 10.1 is this introduction.

Section 10.2 deals with the first one of the aspects said above. Models are discussed such that the state variable of each individual, or object, can attain only a finite number of states. Thus, the evolution equations happen to be simpler than those of the corresponding continuous model since they consist of ordinary differential equations, for models without internal or space structure, or of partial differential equations, for systems with structure. Discrete models are related to the discrete Boltzmann equation [GAa].

Section 10.3 develops some ideas, with reference to [ARb], to design a general framework that may include all, or at least a large variety, of

specific kinetic models. It is shown that this framework also suggests some ideas to develop new classes of models.
Section 10.4 discusses the possibility of constructing models obtained by linking different features that are characteristic of different models such as those presented in the preceding chapters. A more adaptable tool could thus be obtained towards the analysis of large physical systems of interest in applied sciences.

It is obvious that the particular selection of the arguments that conclude this book stems from the authors' personal feelings. Additional topics are certainly of interest. Their identification is left to the reader's initiative; hopefully, the hints given in these Lecture Notes may be used as guidelines for the prosecution of his personal research line.

10.2 Discrete Generalized Models

Discrete models in kinetic theory have been proposed in [GAa] to develop models relatively simpler than the original Boltzmann equation. In addition they should be suitable to provide an immediate description of physical reality and avoid the expensive computational problems related to the corresponding continuous models.

The discrete version of Boltzmann equation, that is the ***discrete Boltzmann equation*** [GAa], refers to a fictitious gas of particles moving in the physical space with only a finite number of velocities. The model is an evolution equation, in particular a system of partial differential equation of hyperbolic type, for the number densities relative to the admissible velocities. The mathematical simplification consists of the fact that a finite set of differential equations replaces the original integrodifferential equation (and its fivefold integration) we have seen in Chapter 7.

A reason to build up discrete models resides in that they easily satisfy natural conservation equations such as conservation of mass, momentum and energy. On the contrary, discretization of the original equation by collocation – interpolation methods, do not generally satisfy the above conservation laws.

It is worth emphasizing that discrete models do not directly correspond to discretization of the original model, but to a suitable simplification (idealization) of the physical phenomenon that is modelled. Therefore, the model validity has to be put in question altogether, and the procedure of suitable comparisons between the real system behavior and the descriptions provided by the model needs to be renewed.

The same lines that lead to the discrete Boltzmann equation may be followed to construct discrete generalized kinetic models. Yet, some other motivations should be provided to support the developing of these last ones. Indeed, the general reasons that we gave here above are not necessarily the same, or the only ones, which are behind the discrete kinetic models.

Discrete Boltzmann equation was proposed mainly for the following reasons:

a) Developing a model, for applications in fluid dynamics, that may replace the original Boltzmann equation when this appeared to be too difficult, or even impossible, to be dealt with.

b) Discussing the qualitative theory of the initial and initial – boundary value problem for a simplified model when the analysis for the Boltzmann equation appeared too hard to be developed.

At present, both the above motivations are no longer valid. As we have seen in Chapter 7, applied mathematicians have provided several interesting results both on the analytic and on the computational treatment of the Boltzmann equation. On the other hand, the analysis of the same problems for the discrete Boltzmann equation gave disappointing results, with the exception of a few solutions to problems in fluid dynamics which still appear to be of interest. Therefore, motivations for developing discrete generalized kinetic models must be found elsewhere.

A reason that we feel sufficiently good resides in the following. Rather than constructing discrete models to simplify the computational and qualitative analysis of the continuous model, it may be of greater interest to develop them with the peculiar aim of providing an identification of the parameters that characterize the model. Indeed this may be done, for the discrete models, in a much faster and more restricted way than for the continuous ones. In particular, highly expensive experiments may be avoided. Hence the results are more understandable and recognizable, and can be achieved with computations that may be more easily corrected and repeated in successive approximating procedures.

The main topic of this section is modelling physical phenomena by means of discrete generalized kinetic models. However, considering that the literature on this topic is very poor, the contents will include suggestions, perspectives and criticisms. The analysis is developed also bearing in mind that the models must be applied to the solution of real problems.

The contents is organized in two subsections. The first one deals with the discrete Boltzmann equation, the second one with the design of a generalized discrete model for the cellular tumors - immune system competition.

Presentation is oriented towards the methodological aspects of the modelling, so that sufficient information is provided to allow the construction of a discrete version of any of the continuous models that we dealt with in the preceding chapters.

10.2.1 The discrete Boltzmann equation

Discrete models refer to a ***fictitious gas*** of particles that can move in space only according to a finite set of velocities $\mathbf{v}_r$, $r = 1, \ldots, m$. The kinetic model is an evolution equation for the number densities

$$N_r \,:\, (t, \mathbf{x}) \in [0, T] \times \Lambda \subseteq \mathbb{R}^3 \;\mapsto\; N_r(t, \mathbf{x}) \in \mathbb{R}_+ \,, \qquad (10.2.1)$$

for $r = 1, \ldots, m$, each one linked to a corresponding value for the velocity. That is, $N_r(t, \mathbf{x})\, d\mathbf{x}$ represents the number of particles in the volume $d\mathbf{x}$ centered at $\mathbf{x}$, that at time t possess velocity $\mathbf{v}_r$.

The idea of discretizing the velocity space belongs to Maxwell, who suggested to split the full range of velocities into two groups of velocities. Despite its extreme simplicity, this idea is still used in stating the boundary conditions for the full Boltzmann equation. In particular, the statement consists in assuming that a certain fraction of the particles that hit the wall at time t and at a certain point of the boundary is instantaneously specularly reflected; the remaining fraction is diffusely re-emitted at the wall temperature. Although this description does not correspond to any physical reality, it is still used as an approximating picture of the phenomenon.

Maxwell's idea was technically developed by Carleman, who proposed a simple two velocities model for particles that can only move along the x-axis. The two velocities have the same modulus; the first velocity refers to particles that move in the positive axis direction, the second in the negative one. Carleman's heuristic model reads

$$\begin{cases} \left(\dfrac{\partial}{\partial t} + \dfrac{\partial}{\partial x}\right) N_+ = \eta(N_-^2 - N_+^2) \\ \left(\dfrac{\partial}{\partial t} + \dfrac{\partial}{\partial x}\right) N_- = \eta(N_+^2 - N_-^2)\,, \end{cases} \tag{10.2.2}$$

where N_+ and N_- are defined on $[0, T] \times \mathbb{R}$ and have values in $\mathbb{R}_+$.

In fact, it is hardly possible to relate Carleman's model to the Boltzmann equation. Not only because the first one is too simple, but also because its derivation does not even follow similar rules and, in particular, similar conservation equations. All the same it can be regarded as an example of a heuristic derivation of a simple two-velocity model. In the following, when we talk about discrete Boltzmann equation we shall assume that the number m of velocities is significantly large, and that the interaction between gas particles preserves mass, momentum, and energy.

The general assumptions that lead to the discrete Boltzmann equation will now be given together with the model's formal structure. It is worth stressing, however, that the following general assumptions lead to a whole class of models, and not to a unique one. Each particular example can then be derived under additional detailed assumptions on velocity discretization and particle interactions.

Assumption 10.2.1. *The system is constituted by a large number N of not distinguishable interacting particles of unitary mass and (isotropic, uniform) cross sectional area $\mathcal{B}$. Particles are supposed to move with velocities that may take only a finite number of values from a set*

$$I_{\mathbf{v}} = \{\mathbf{v}_r\}_{r=1}^{m}, \qquad \mathbf{v}_r \in \mathbb{R}^3, \quad r = 1, \ldots, m. \tag{10.2.3}$$

The statistical distribution of the system is identified by the family $\mathbf{N} := \{N_r\}_{r=1}^{m}$ of the number density functions N_r defined in (10.2.1).

Assumption 10.2.2. *Collisions between particles, with pre-collision velocities $\mathbf{v}_i$, $\mathbf{v}_j$ and post-collision velocities $\mathbf{v}_h$, $\mathbf{v}_k$, are instantaneous, localized in space, and such that mass, momentum, and energy are preserved*

$$\begin{cases} \mathbf{v}_i + \mathbf{v}_j = \mathbf{v}_h + \mathbf{v}_k, \\ v_i^2 + v_j^2 = v_h^2 + v_k^2, \end{cases} \qquad i, j, h, k \in \{1, \ldots, m\}. \tag{10.2.4}$$

Assumption 10.2.3. *Only binary collisions are taken into account. The probability of higher order collisions is taken to be zero. Particles cannot be individually recognized. Collisions are reversible.*

Assumption 10.2.4. *The expected number of collisions between particles with incoming velocities $\mathbf{v}_i$ and $\mathbf{v}_j$, that happen at times in dt centered at t and places in $d\mathbf{x}$ centered at $\mathbf{x}$, is*

$$\mathcal{B}\, |\mathbf{v}_i - \mathbf{v}_j|\, N_i(t, \mathbf{x})\, N_j(t, \mathbf{x})\, dt\, d\mathbf{x}, \qquad i, j \in \{1, \ldots, m\}. \tag{10.2.5}$$

Clearly, the same number is also given by

$$\mathcal{B}\, |\mathbf{v}_h - \mathbf{v}_k|\; N_h(t,\mathbf{x})\; N_k(t,\mathbf{x})\; dt\, d\mathbf{x}\,, \qquad h,k \in \{1,\ldots,m\}\,. \tag{10.2.6}$$

Assumption 10.2.5. *Collision dynamics identifies a* ***transition probability distribution*** $\{\psi_{ij}^{hk}\}_{i,j,h,k}^{1,\ldots,m}$. *The term* $\psi_{ij}^{hk} \in \mathbb{R}_+$ *defines the (conditional) probability that an encounter between a pair of particles with velocities* $\mathbf{v}_i$ *and* $\mathbf{v}_j$ *ends up in the (same) particles with velocities* $\mathbf{v}_h$ *and* $\mathbf{v}_k$. *The above probabilities are modelled by taking into account that admissible collisions must preserve mass, momentum, energy, and be consistent with the set* $I_\mathbf{v}$. *The normalization and reversibility properties are also satisfied:*

$$\sum_{h,k=1}^{m} \psi_{ij}^{hk} = 1\,, \qquad i,j \in \{1,\ldots,m\}\,, \tag{10.2.7}$$

and, for all $i,j,h,k \in \{1,\ldots,m\}$,

$$\psi_{ij}^{hk} = \psi_{ji}^{hk} = \psi_{ij}^{kh} = \psi_{hk}^{ij} = \psi_{hk}^{ji} = \psi_{kh}^{ij}\,. \tag{10.2.8}$$

A particular choice is:

$$\psi_{ij}^{hk} = \frac{1}{\nu(i,j;h,k)}\,, \qquad i,j,h,k \in \{1,\ldots,m\}\,, \tag{10.2.9}$$

where $\nu(i,j;h,k)$ *denotes the number of admissible velocity outputs* h,k *at given inputs* i,j.

Similarly to the continuous model, the phenomenologic derivation of discrete Boltzmann equations is obtained by the balance equations

$$\left(\frac{\partial}{\partial t} + \mathbf{v}_i \frac{\partial}{\partial \mathbf{x}}\right) N_i = G_i[\mathbf{N}] - L_i[\mathbf{N}]\,, \qquad i = 1,\ldots,m\,, \tag{10.2.10}$$

where $\mathbf{N} = \{N_i\}_{i=1}^m$, and where the ***gain*** and ***loss*** terms are given by

$$G_i[\mathbf{N}](t,\mathbf{x}) = \frac{1}{2}\sum_{jhk}^{1,m} \mathcal{B}\ |\mathbf{v}_h - \mathbf{v}_k|\ \psi_{ij}^{hk}\ N_h(t,\mathbf{x})\ N_k(t,\mathbf{x})\,, \qquad (10.2.11)$$

and

$$L_i[\mathbf{N}](t,\mathbf{x}) = \frac{1}{2}\sum_{jhk}^{1,m} \mathcal{B}\ |\mathbf{v}_i - \mathbf{v}_j|\ \psi_{ij}^{hk}\ N_i(t,\mathbf{x})\ N_j(t,\mathbf{x})\,. \qquad (10.2.12)$$

Hence, the general expression of the discrete Boltzmann equation for binary collisions may be summarized as follows

$$\left(\frac{\partial}{\partial t} + \mathbf{v}_i\frac{\partial}{\partial \mathbf{x}}\right) N_i = \frac{1}{2}\sum_{jhk}^{1,m} J_{ij}^{hk}\,(N_h N_k - N_i N_j)\,. \qquad (10.2.13)$$

Mathematical problems are stated, as for the continuous Boltzmann equation, linking the evolution equation to suitable initial and (or) boundary conditions. A treatment of the boundary conditions is given in [GAb]. Generalizations to multiple collisions are possible as documented in [BEa]. The vast literature on the above model is not cited here, but readers can refer to two review papers [PLa] and [BEa], where the pertinent literature can be recovered. The first paper deals mainly with the derivation and possible applications of the model; the second one is mathematically oriented to the analytic treatment of the evolution problems. Additional reference is made to the Lecture Notes by Gatignol [GAa], where a detailed analysis of the derivation of the above model and of its thermodynamic properties can be found, and to the book [MOa], where several applications in fluid dynamics are described. Specific models, corresponding to particular discretization schemes, are reported in the cited review papers and books.

10.2.2 Discrete models in immunology

In this subsection the derivation of discrete state models are dealt with in relation with the generalized kinetic models for immune system – tumor cells competition. The corresponding continuous cases have been described in Chapter 5. Here, the discretization is developed with reference to [LOa].

The aim is tutorial, i.e., it offers the reader a methodology that may be used as a basis to design discrete models somewhat related to analogous continuous ones.

Motivations to develop the following discrete models mainly reside in obtaining a class of mathematical problems whose solution is relatively simpler, and certainly faster to compute, than those of the continuous ones, and whose related parameters are more easily identifiable. In fact, even if an oriented reading may further simplify the continuous model, and show that only some of the interactions are indeed effective for the system, nonetheless taking advantage of the experimental evidence is a hard task. It becomes even more difficult when one has to define the correct, or most probable, functional forms of the various kernels that appear in integral terms of the continuous model.

Conversely, although discretization generally yields only a rough approximation of the observed phenomena, it provides an immediate description of the system behavior, since it introduces only a finite set of control parameters. For instance, as seen above, probability density functions reduce to finite sets of positive numbers. Moreover, the discrete assumption on the state variable of a generalized kinetic model appears to be less striking than discretizing the velocity variable in a strictly kinetic model. Indeed, allowing a finite number of activity values, for the cells that are involved in the immune cells – tumor competition, may be *a priori* as meaningful and acceptable as the corresponding continuum assumption.

For the following models the notations and starting points are those of Chapter 5. In it, n is the number of species involved, and each species activity takes values in a discrete set:

$$I_u := \{u_1 = -1, \ldots, u_k \ldots, u_m = +1\}\,. \qquad (10.2.14)$$

Hence the number densities, that constitute the unknowns of the problem, are

$$N_{ir} = N_{ir}(t)\ , \qquad i = 1, 2, \ldots, n\ , \qquad r = 1, 2, \ldots, m\,, \quad (10.2.15)$$

meaning that $N_{ir}(t)$ represents the number of particles of the i-th population that at time t are in the r-th activation state. They are densities in that the total number of particles of i-th population is given by

$$N_i(t) = \sum_{r=1}^{m} N_{ir}(t)\,, \qquad i = 1, \ldots, n\,. \qquad (10.2.16)$$

It is worthwhile stressing again that this discretization is not just a stepwise numerical subdivision of the range of a continuous variable, a necessary procedure to compute its continuous behavior, but rather an *a priori* drastic reduction of its range to a fixed discrete set. In this way the characteristic states may be represented only at large, but the model has a number of steps that is strongly reduced. For instance, computations on the particular model that follows, and that all the same correctly describes the qualitative properties of the physical system, have been realized for $m = 2$.

A dynamical system that somehow corresponds to Eq. (5.4.2) is obtained by replacing the various continuous densities by a corresponding discrete set of probabilities. This produces a model which, in fact, may be considered as only a special case of any other model of population dynamics with $n \cdot m$ different species. Yet, it is not only more understandable, since the effect of social rules and control parameters are more clearly assigned and interpreted, but also it straightforwardly allows to increase the accuracy by increasing the number m without reconsidering all the details.

In addition, the resulting system of ordinary differential equations inherits a proper interest by its being a derivation of a generalized kinetic model, i.e., a set of integrodifferential equations involving second order products of probability density functions and subject to appropriate conservation laws. In particular, the analysis may be relevant to those consequences on the dynamics that directly stem from the special relations on the coefficients due to corresponding bounds among various terms of the continuous model.

Necessarily, the discrete model must have the following characteristic features: its dynamics has to depend on the activation state; it must contain the competition among the various cell populations; it may depend on a relatively small number of control parameters; its results should be directly comparable with the experimental evidence; its parameters should be put in a direct correspondence with the observed chemical and physical variables that rule the system.

The simplest, although general, system of ordinary differential equations which fulfills all these requirements, and that may represent a kinetic discrete model exhibiting both conservative and proliferative events, has $m = 2$ and $n = 3$. This will be assumed in the following, with some even further specialization based on a modelling devoted to the case of tumor-immune system cells competition.

Yet, in spite of all these assumptions, the resulting system of equations depends on a threatening set of 35 real parameters to be identified. The unexpectedly easy discussion which happens to be possible in this particular case proves to be a merit of the kinetic origin of the model.

Bearing this in mind, call I_n, I_m two sets of n and m symbols respectively, and let $N_{ir} = N_{ir}(t)$ be the population densities of species $i \in I_n$ in the activation state $r \in I_m$. The model may be summarized at first by the following $n \cdot m$ ordinary differential equations:

$$\frac{d}{dt}N_{ir} = \sum_{\substack{j \in I_n \\ h,k \in I_m}} a(ih, jk; r)\; N_{ih}N_{jk} + \sum_{\substack{j \in I_n \\ h,k \in I_m}} b(ih, jk; r)\; N_{ih}N_{jk}\,, \qquad (10.2.17)$$

for $i \in I_n$ and $r \in I_m$, and where conservative and proliferative terms have been grouped, and the corresponding coefficients defined as follows.

Conservative interactions are ruled by the coefficients $a(ih, jk; r)$, for $i, j \in I_n$, and $h, k, r \in I_m$, that are product of a kinetic term and a probabilistic one according to:

$$\begin{aligned} a(ih, jk, r) \;&:= \eta(ih, jk)\;\left(\psi(ih, jk; r)\; - \delta_{hr}\right) \\ &:= \eta_{ij}(v_h, w_k)\left(\psi_{ij}(v_h, w_k; u_r) - \delta(v_h - u_r)\right)\,. \end{aligned} \qquad (10.2.18)$$

The ***encounter rate*** $\eta(ih, jk)$ is the number, per unit volume and unit time, of encounters between a cell of population i in state h and a cell of population j in state k. The ***transition coefficient*** $(\psi(ih, jk; r)\; - \delta_{hr})$ contains the ***transition probability*** $\psi(ih, jk; r)$ about the event that an i-cell in state h is transformed into an i-cell in state r due to an encounter with a j-cell in state k, and it satisfies

$$\sum_{r \in I_m} \psi(ih, jk; r) = 1\,, \qquad i, j \in I_n, \quad h, k \in I_m\,. \qquad (10.2.19)$$

The transition coefficient is normalized by the Kronecker symbols δ_{hr} due to the conservative character of these interactions which are such that individuals of a certain population may change state only if the total number of cells of that population remains constant.

Proliferative and ***Destructive interactions*** are ruled by the terms $b(ih, jk; r)$, for $i, j \in I_n$, and $h, k, r \in I_m$, that are the difference between the proliferative and destructive parts according to:

$$\begin{aligned} b(ih,jk;r) &:= p(ih,jk)\,\varphi(ih,jk;r) \;-\; d(ih,jk) \\ &:= p_{ij}(v_h,w_k)\varphi_{ij}(v_h,w_k;u_r) - d_{ij}(v_h,w_k)\,. \end{aligned} \tag{10.2.20}$$

A cell may proliferate only into a cell of the same species, yet in any of the m possible activity states. Proliferation is ruled by a kinetic coefficient and a probabilistic one. The ***fertility rate*** $p(ih,jk)$ denotes the total number of cells per unit volume and unit time that a cell of population i in state h is able to produce because of an encounter with a cell of population j in state k. The ***proliferative probability*** $\varphi(ih,jk;r)$ denotes the fraction, of the $p(ih,jk)$ cells, that is produced in the r-th activity state; and it satisfies

$$\sum_{r\in I_m} \varphi(ih,jk;r) = 1\,, \qquad i,j\in I_n, \quad h,k\in I_m\,. \tag{10.2.21}$$

The term $d(ih,jk)$ controls the ***destruction rate*** of species i in state h due to an interaction with j-cells in state k.

Neither proliferative nor destructive terms relative to a certain species i have any connections with the same terms relative to different species.

The system of ordinary differential equations which represents a general discrete model that exhibits, both and separately, conservative and proliferative events may then be written as follows, for $i\in I_n$, and $r\in I_m$,

$$\begin{aligned} \frac{d}{dt}N_{ir} = &\sum_{\substack{j\in I_n\\ h,k\in I_m}} \eta(ih,jk)\,(\psi(ih,jk;r)-\delta_{hr})\,N_{ih}N_{jk} \\ &+ \sum_{\substack{j\in I_n\\ h,k\in I_m}} (p(ih,jk)\varphi(ih,jk;r)-d(ih,jk))\,N_{ih}N_{jk}\,. \end{aligned} \tag{10.2.22}$$

As already mentioned, we now restrict Eq. (10.2.22) to the special case of the tumor-immune cells dynamics in a medium of inert cells. The population of the medium is assumed to be so numerous that it can be considered as a constant, and hence the medium may have just one activation state. This leads to $n = 3$ and $m = 2$. We shall denote by

$N_{11} =: T_a$, the active tumor population,

$N_{12} =: T_p$, the passive tumor population,

$N_{22} =: Q_p$, the passive immune-cells population,

$N_{21} =: Q_a$, the active immune-cells population,

$N_{31} =: M$, the inert cells population .

All the variables are assumed to be non-negative continuous functions of time, defined over a closed interval $[t_0, t_1] \subset \mathbb{R}$.

As it is done in the continuous case, and *a fortiori* here where the aim is to produce an easily computable model, some further phenomenologic assumptions are now necessary to lower the number of the undetermined parameters that still appear in the resulting equations. These assumptions are consequences of detailed observations of the physical system at the microscopic scale and, clearly, need an *a posteriori* confirmation based on actual observations and simulated results.

The conservative terms of the model are subject to the following special:

Biological Rules I

a) Immune cells activity is not raised by encounters with active tumors.

Immune cells activity is not lowered by encounters with passive tumors.

Immune cells do not change their activity values when they encounter inert cells.

b) Tumor cells activity is not raised by encounters with active immune cells.
Tumor cells activity is not lowered by encounters with passive immune cells.
Tumor cells do not change their activity values when they encounter inert cells.

c) No species changes its activity on encountering cells of the same species.

Consequently, the following are the only conservative terms which still appear in the dynamical system

$$\eta(T_p, Q_p) = \alpha_1\,, \qquad \eta(T_a, Q_a) = \alpha_4\,,$$

and

$$\psi(T_p, Q_p; T_p) = 1 - \tau_1\,, \qquad \psi(T_p, Q_p; T_a) = \tau_1\,,$$

$$\psi(T_a, Q_a; T_a) = 1 - \tau_2\,, \qquad \psi(T_a, Q_a; T_p) = \tau_2\,,$$

$$\psi(Q_a, T_a; Q_a) = 1 - \tau_3\,, \qquad \psi(Q_a, T_a; Q_p) = \tau_3\,,$$

$$\psi(Q_p, T_p; Q_p) = 1 - \tau_4\,, \qquad \psi(Q_p, T_p; Q_a) = \tau_4\,,$$

where $\alpha_1, \alpha_4 \geq 0, \quad 0 \leq \tau_i \leq 1, \quad i = 1, \ldots, 4.$

Furthermore, proliferative and destructive terms of the model are subject to the following

Biological Rules II

d) Tumors may perish only when encountering immune cells.
Immune cells ***naturally*** extinguish, i.e., on encounters with inert cells. This behavior may possibly happen on a different time scale, and it may be further corrected by taking into account, in form of average, the compensating effect due to bone marrow cells production.

e) Tumors may proliferate not only on encounters with other tumors or inert cells, but also with passive immune cells.
Immune cells proliferate only if tumors are present.
Each species may proliferate only in cells of the same species.

This second set of rules gives the following coefficients

$$d(T_a, Q_a) = -\delta_1\,, \qquad d(T_a, Q_p) = 0\,,$$
$$d(T_p, Q_a) = -\delta_2\,, \qquad d(T_p, Q_p) = -\delta_4\,,$$
$$d(Q_a, M) = -\delta_3\,, \qquad d(Q_p, M) = -\delta_5\,,$$

where $\delta_i \geq 0, \quad i = 1, \ldots, 5,$

$$p(T_a, M) = \nu_1\,, \qquad p(T_p, T_a) = \nu_5\,, \qquad p(Q_p, T_a) = \mu_1\,,$$
$$p(T_a, T_p) = \nu_2\,, \qquad p(T_p, T_p) = \nu_6\,, \qquad p(Q_p, T_p) = \mu_2\,,$$
$$p(T_a, T_a) = \nu_3\,, \qquad p(T_a, Q_p) = \nu_7\,, \qquad p(Q_a, T_a) = \mu_3\,,$$
$$p(T_p, M) = \nu_4\,, \qquad p(Q_a, T_p) = \mu_4\,,$$

where $\nu_i \geq 0, \quad i = 1, \ldots, 7, \quad \mu_i \geq 0, \quad i = 1, \ldots, 4,$ and

$$\varphi(T_a, M; T_a) = 1 - \gamma_1\,, \qquad \varphi(T_a, M; T_p) = \gamma_1\,,$$
$$\varphi(T_a, T_p; T_a) = 1 - \gamma_2\,, \qquad \varphi(T_a, T_p; T_p) = \gamma_2\,,$$
$$\varphi(T_a, T_a; T_a) = 1 - \gamma_3\,, \qquad \varphi(T_a, T_a; T_p) = \gamma_3\,,$$
$$\varphi(T_p, M; T_p) = 1 - \gamma_4\,, \qquad \varphi(T_p, M; T_a) = \gamma_4\,,$$
$$\varphi(T_p, T_a; T_p) = 1 - \gamma_5\,, \qquad \varphi(T_p, T_a; T_a) = \gamma_5\,,$$
$$\varphi(T_p, T_p; T_p) = 1 - \gamma_6\,, \qquad \varphi(T_p, T_p; T_a) = \gamma_6\,,$$
$$\varphi(T_a, Q_p; T_a) = 1 - \gamma_7\,, \qquad \varphi(T_a, Q_p; T_p) = \gamma_7\,,$$

where $0 \leq \gamma_i, \quad i = 1, \ldots, 7,$

$$\varphi(Q_p, T_a; Q_a) = \beta_1\,, \qquad \varphi(Q_p, T_a; Q_p) = 1 - \beta_1\,,$$
$$\varphi(Q_p, T_p; Q_a) = \beta_2\,, \qquad \varphi(Q_p, T_p; Q_p) = 1 - \beta_2\,,$$
$$\varphi(Q_a, T_a; Q_p) = \beta_3\,, \qquad \varphi(Q_a, T_a; Q_a) = 1 - \beta_3\,,$$
$$\varphi(Q_a, T_p; Q_p) = \beta_4\,, \qquad \varphi(Q_a, T_p; Q_a) = 1 - \beta_4\,,$$

where $0 \leq \beta_i \leq 1, \quad i = 1, \ldots, 4$.

In addition, the constants γ_a and γ_p are considered as consequences of the cell production by the bone-marrow.

Although discretization has drastically reduced the dimension of the problem by limiting the number of the population densities to be described, anyway if the system of ordinary differential equations is explicitly written at this stage then it is found to depend on a still uneasy number of control parameters.

The last step in the model reduction consists in establishing a set of reasonable, although somewhat unjustified, estimates on some of the 35 a priori constants $(\alpha, \tau, \nu, \mu, \gamma, \beta, \delta)$, that will lead to a ***first approximation*** system of equations. Hopefully this last system only depends on the characteristic, meaningful, control parameters that may be put in direct correspondence with the physico-chemical variables.

First-approximation Assumptions

a) Fertility of tumor cells in encounters with other tumor cells is negligible with respect to that in encounters with inert or passive immune cells. Fertility of passive tumors is much lower than that of active ones; (i.e., $\nu_2 = \nu_3 = \nu_5 = \nu_6 = 0, \quad \nu_4 = \nu_1 \epsilon, \quad \epsilon \ll 1$).

b) Proliferation of tumor cells in encounters with inert or passive immune cells produces tumors mostly in the same activity state rather than in different ones; (i.e., $\gamma_1 = \gamma_4 = \gamma_7 = \epsilon \ll 1$).

c) Fertility of immune cells in encounters with active tumor cells is much higher than that with passive ones. Moreover it has the same values for active as well as for passive immune cells; (i.e., $\mu := \mu_1 = \mu_3 \gg \mu_2 = \mu_4 = \mu\epsilon$).

d) Activity state of new born immune cells is almost certainly the same as that of fathers' if and only if the tumor cells that conditioned the interaction were in passive status. That is: on encounters with passive tumors, immune cells proliferate in immune cells of the same activity. On encounters with active tumors their change of state is frequent and the same for both active and passive immune cells; (i.e., $\beta := \beta_1 = \beta_3 \gg \beta_2 = \beta_4 = \epsilon$). In all cases, proliferation in cells of the same state is more probable than in different ones; (i.e., $1/2 \leq \beta \leq 1$).

e) Mortality of passive tumor cells does not depend on the activity of immune cells responsible for their death; (i.e., $\delta_2 = \delta_4$).

f) On encounters with passive tumors, passive immune cells do not have enough energy to rise to their active state; (i.e., $\tau_4 = 0$).

This set of assumptions finally yields, at first order in ϵ, to the following dynamical system, which consists of four highly nonlinear ordinary differential equations depending on twelve constitutive (real) parameters:

$$\begin{cases} \dfrac{d}{dt}T_a = +\nu_1(1-\epsilon)T_aM + \nu_7(1-\epsilon)T_aQ_p \\ \qquad -(\delta_1 + \alpha_4\tau_2)T_aQ_a + \alpha_1\tau_1T_pQ_p \\ \dfrac{d}{dt}T_p = +\nu_1\epsilon T_pM + \nu_7\epsilon T_aQ_p + \alpha_4\tau_2T_aQ_a \\ \qquad -(\delta_2 + \alpha_1\tau_1)T_pQ_p - \delta_2T_pQ_a + \nu_1\epsilon T_aM \\ \dfrac{d}{dt}Q_p = -\,\delta_5Q_pM + \mu(1-\beta)T_aQ_p \\ \qquad +(\mu\beta + \alpha_4\tau_3)T_aQ_a + \mu\epsilon T_pQ_p + \gamma_p \\ \dfrac{d}{dt}Q_a = -\,\delta_3Q_aM + \mu\beta T_aQ_p \\ \qquad +(\mu(1-\beta) - \alpha_4\tau_3)T_aQ_a + \mu\epsilon T_pQ_a + \gamma_a\,. \end{cases} \tag{10.2.23}$$

Again we point out that Eq. (10.2.23) is in fact a model for a physical system only if the parameters α, ν, δ, γ, and μ are positive real quantities, τ and β in $[0,1]$, and ϵ is a positive number much less than one.

The model represented by Eq. (10.2.23) can be conveniently discussed both with respect to its solutions dependence from initial conditions and to its structural stability on the parameters. Clearly, owing to the dimensionality of the parameters space, the discussion is involved and is far from being an exhaustive character. Anyway, it is not too difficult to realize that some of the parameters play a role on the system dynamics which is preeminent with respect to the others. For instance it is immediate to see that the proliferation constants ν_1, ν_7, transition constants α_1, α_4, and particularly the fertility coefficient μ are critical quantities. Conversely, the ***small*** terms $\epsilon\nu_1 T_p$, $\epsilon\mu T_p Q_p$, $\epsilon\mu T_p Q_a$, actually produce consequences which are irrelevant on the (asymptotic) dynamic, and thus they may be discarded in a first approximation analysis.

Similarly, a preeminent role is played by the constants δ_3, δ_5 and γ_a, γ_p which control the unconditioned production-extinction of immune cells. These parameters give rise to a bifurcating behavior. When the gamma's are not negligible, immune cells (and consequently tumor cells) may be asymptotically driven towards a set of constant values that depend on the gamma's and that, if the gamma's are small enough, are reached with a kind of damping. However, this behavior is connected with the pair (δ_3, δ_5) in such a way that if $\delta_3 < \delta_5$, and if they are both sufficiently great with respect to the gamma's, then the system may diverge; conversely, it collapses if $\delta_3 > \delta_5$.

On the basis of these remarks, the analysis may be separated according to the values of constants δ_3, δ_5 and γ_a, γ_p. In [LOa] the more interesting case has been exploited of $\gamma_a = \gamma_p = 0$, $\delta_3 = \delta_5 =: \delta$, and non-trivial values for the remaining constants. Further examination of the behavior of the system is of interest.

Unexpectedly, the kinetic origin of the model together with all these assumptions produce a kind of retarded dependence of a pair of variables from the other pair. Indeed, after a short transient time, the values of T_a and Q_a become proportional to those of T_p and Q_p respectively. This may be explicitly seen by means of a trivial change of variables:

$$Q_a = x\ , \quad Q_p = cx + \xi\ , \quad T_a = y\ , \quad T_p = ky + \eta\ , \qquad (10.2.24)$$

where

$$c = \left(1 + \frac{\alpha_4 \tau_3}{\mu\beta}\right)\ , \qquad \mu\beta \neq 0\ , \qquad (10.2.25)$$

and where k depends on various parameters of the problem in that it is a solution of certain algebraic equation of the second order.

The above change of variables (and the mentioned approximating assumptions) transform Eq. (10.2.23) into

$$\begin{cases} \dfrac{d\xi}{dt} = -\delta\xi - c_1\xi y \\ \dfrac{dx}{dt} = -\delta x + \mu xy + \mu\beta\xi y \\ \dfrac{d\eta}{dt} = -(b_1 + c^{-1}b_2\xi)y - \delta_2 x\eta - b_3(cx + \xi)\eta \\ \dfrac{dy}{dt} = +\nu_1(1-\epsilon)y - a_1 xy + \alpha_1\tau_1(cx + \xi)\eta + a_2\xi y\ , \end{cases} \qquad (10.2.26)$$

where constants $a_1, a_2, b_1, b_2, b_3, c_1$ have been introduced conveniently depending on the preceding ones. In fact, the change of variable (10.2.24), and the equivalence between Eq. (10.2.26) and Eq. (10.2. 23), depend on a further combination:

$$a_3 := +\delta_1 + \alpha_4\tau_2 - c\nu_7(1-\epsilon) - \delta_2(1+c)$$

of the parameters above, on behalf of which the constant k may be found, positive.

It is immediate to see that whatever the values of the four initial conditions and x, y, η are, the values $\xi(t)$ rapidly tend to zero. Simulation shows that the decay time is very short, a few time-units, and this characterizes the transient.

A similar behavior may happen to the variable η under the additional condition $a_3 < 0$. The asymptotic value of η critically depends on a combined balance of the parameters that also define the system equilibrium point, which is

$$(x,\, y,\, \xi,\, \eta)_{eq} = \left(\frac{b_5}{a_1(\delta_2 + b_3 c)},\, \frac{\delta}{\mu},\, 0,\, -\frac{a_1 b_1 \delta}{b_5 \mu} \right), \qquad (10.2.27)$$

where

$$b_5 := \nu_1(1-\epsilon)(\delta_2 + b_3 c) - c_1 \alpha_1 \tau_1 b_1 .$$

It is now clear that the main bifurcating situations are:

$$a_1 = 0\,, \quad a_3 = 0\,, \quad b_5 = 0\,, \qquad (10.2.28)$$

which point out the relevance of some of the parameters with respect to the others and hence provide hints on their comparison with the experimental controls and therapies.

Also from the mathematical point of view it is remarkable that, when $a_3 < 0$, the system fast collapses on the two-dimensional phase space of the variables T_a and Q_a on which its behavior is ruled by the extremely simple and classical predator-prey two population equation:

$$\begin{cases} \dfrac{dx}{dt} = -\delta x + \mu x y\,, \\[2mm] \dfrac{dy}{dt} = +\nu_1(1-\epsilon) y - a_1 x y\,. \end{cases} \qquad (10.2.29)$$

It consequently follows that the discrete model, although so reduced by all the various assumptions and approximations we mentioned so far, shows a behavior in complete agreement with that evaluated by the correspondent continuous model, see e.g., [BEd], [FIa].

10.3 Looking for a General Structure

As we have seen in the preceding chapters, and also discussed in Chapter 1, the mathematical structure of generalized kinetic models is not the same for all classes of models. Therefore, in principle, one cannot provide a unique formal structure that may be the same for all the classes of generalized kinetic models.

However, following some ideas developed in [ARc], we will now describe a general framework for a large variety of models that may simulate the evolution of a great number of individuals, or objects, undergoing kinetic interactions and competitions. This will not only provide a general picture suitable to generate specific models for the applications, but may also address research activity towards the development of new models.

Consider a system of several interacting populations of anonymous individuals or objects, undistinguishable within each population. Each individual is characterized by a certain ***state***, each population by a certain ***size***, i.e., by the number of its individuals. The system evolution is determined by interactions between pairs of individuals. Interactions happen at the microscopic scale, and may modify the distribution functions over the states and/or the sizes.

Here we deal with the relatively simpler case of models with a behavior homogeneous with respect to the space variable. Then further generalizations to models with time and space structure can be given. The class of models is described by the following assumptions:

Assumption 10.3.1. *The system consists of n interacting populations. Each individual of a population is characterized by the **state** $\mathbf{u} \in I \subseteq \mathbb{R}^m$, where $m \in \mathbb{N}$. If the components u_i of $\mathbf{u}$ are defined on bounded intervals, the range of $\mathbf{u}$ is assumed to be $\mathbf{I} = [-1,1]^m$.*

Assumption 10.3.2. *The populations number densities with respect to the state variable $\mathbf{u}$, at time t, are defined by*

$$f_i \,:\, (t,\mathbf{u}) \in [0,T] \times \mathbf{I} \;\mapsto\; f_i(t,\mathbf{u}) \in \mathbb{R}_+\,, \qquad i = 1,\ldots,n\,.$$

The number of individuals, i.e., the ***size****, of i-th population at time* $t \in [0, T]$ *is identified by*

$$N_i(t) \equiv E_0[f_i](t) = \int_{\mathbf{I}} f_i(t, \mathbf{u})\, d\mathbf{u}\,. \tag{10.3.1}$$

The total number of individuals in the system at time t is

$$N(t) = \sum_{i=1}^{n} N_i(t)\,. \tag{10.3.2}$$

Assumption 10.3.3. *Interactions between pairs of individuals are homogeneous in space and instantaneous, i.e., without a delay time, and may modify both size and state; and manufacture individuals of other populations. Only binary encounters are significant to the evolution of the system.*

Assumption 10.3.4. *Modelling the interaction rate between pairs of individuals, of the same or different populations, is based on the identification of the* ***encounter rate***

$$\eta_{jk} : (\mathbf{v}, \mathbf{w}) \in \mathbf{I} \times I \;\mapsto\; \eta_{jk}(\mathbf{v}, \mathbf{w}) \in \mathbf{R}_+\,, \tag{10.3.3}$$

that is the number of encounters, per unit time, of individuals of the j-th population in state $\mathbf{v}$ *with individuals of the k-th population in state* $\mathbf{w}$.

Assumption 10.3.5. *Modelling the state modifications of interacting individuals, of the same or different populations, is based on the identification of the* ***interaction-transition function***

$$g_{jk}^{(i)} : (\mathbf{v}, \mathbf{w}; \mathbf{u}) \in \mathbf{I} \times \mathbf{I} \times \mathbf{I} \;\mapsto\; g_{jk}^{(i)}(\mathbf{v}, \mathbf{w}; \mathbf{u}) \in \mathbb{R}_+\,, \tag{10.3.4}$$

that is the (conditional) probability that individuals of the i-th population in state $\mathbf{u}$ *are manufactured when individuals of the j-th-population in state* $\mathbf{v}$ *interact with individuals of the k-th population in state* $\mathbf{w}$. *The product between η and g is the* ***interaction-transition rate***

$$\eta_{jk}(\mathbf{v},\mathbf{w})\, g_{jk}^{(i)}(\mathbf{v},\mathbf{w};\mathbf{u})\,. \tag{10.3.5}$$

The model consists in a set of evolution equation for the number densities

$$\mathbf{f} = \{f_i\}_{i=1}^{p}; \quad f_i\,:\,(t,\mathbf{u}) \in [0,T]\times \mathbf{I} \;\mapsto\; f_i(t,\mathbf{u}) \in \mathbb{R}_+\,. \tag{10.3.6}$$

The evolution equation for the density f_i can be derived by a balance which equates the time derivative of f_i to the difference between the ***gain*** term G_i and the ***loss*** term L_i. The gain term models the rate of increase of the distribution function due to individuals which fall into state $\mathbf{u}$ of the i-th population due to pair uncorrelated interactions. The loss term models the rate of loss of individuals from their state $\mathbf{u}$, or from the i-th population, due to pair interactions. The evolution equation reads

$$\frac{\partial f_i}{\partial t} = J_i[\mathbf{f}] = G_i[\mathbf{f}] - L_i[\mathbf{f}]\,, \tag{10.3.7}$$

for $i = 1,\ldots,n$, and where

$$G_i[\mathbf{f}](t,\mathbf{v}) = \sum_{j,k=1}^{n} \int_{\mathbf{I}\times\mathbf{I}} \eta_{jk}(\mathbf{v},\mathbf{w}) g_{jk}^{(i)}(\mathbf{v},\mathbf{w};\mathbf{u}) f_j(t,\mathbf{v})\; f_k(t,\mathbf{w})\, d\mathbf{v}\, d\mathbf{w}\,, \tag{10.3.8a}$$

and

$$L_i[\mathbf{f}](t,\mathbf{v}) = f_i(t,\mathbf{u}) \sum_{j=1}^{n} \int_{\mathbf{I}} \eta_{ij}(\mathbf{u},\mathbf{v}) f_j(t,\mathbf{v})\, d\mathbf{v}\,. \tag{10.3.8b}$$

Formally integrating Eq. (10.3.7) with respect to $\mathbf{u} \in \mathbf{I}$ yields

$$\frac{dN_i}{dt}(t) = \sum_{j,k=1}^{n} \int_{\mathbf{I}\times\mathbf{I}} J_{jk}^{(i)}[\mathbf{f}](t,\mathbf{v},\mathbf{w})\, d\mathbf{v}\, d\mathbf{w}\,, \qquad (10.3.9)$$

where

$$J_{jk}^{(i)}[\mathbf{f}](t,\mathbf{v},\mathbf{w}) = \left(\int_{\mathbf{I}} g_{jk}^{(i)}(\mathbf{v},\mathbf{w};\mathbf{u})\, d\mathbf{u} - \delta_{ij}\right)$$

$$\times\ \eta_{jk}(\mathbf{v},\mathbf{w})\, f_j(t,\mathbf{v})\, f_k(t,\mathbf{w})\,, \qquad (10.3.10)$$

and where δ_{ik} is the Kronecker symbol. If all the η_{jk} and $g_{jk}^{(i)}$ are constants, then Eq. (10.3.9) takes the form

$$\frac{dN_i}{dt} = |\mathbf{I}| \sum_{j,k=1}^{n} \eta_{jk}\, g_{jk}^{(i)}\, N_j\, N_k \; - N_i \sum_{k=1}^{n} \eta_{ik} N_k \qquad (10.3.11)$$

where $|\mathbf{I}|$ is the measure of $\mathbf{I}$, and where each of the N_i's is a function from $[0,T]$ to $\mathbf{R}_+$.

A further generalization of the model can be developed by inserting in the evolution equation a production, or ***migration term*** $\gamma_i = \gamma_i(t,\mathbf{u})$ of individuals of the i-th population to state $\mathbf{u}$ due to any artificial inlet. In this case the term

$$\Gamma_i(t) = \int_{\mathbf{I}} \gamma_i(t,\mathbf{u})\, d\mathbf{u} \qquad (10.3.12)$$

has to be added to the right-hand side of Eq. (10.3.9).

Referring to models with zero source term: $\gamma_i = 0$, for $i = 1,\ldots,n$, a preliminary classification can be organized on the basis of the value of the quantity

$$G_{jk}(\mathbf{v},\mathbf{w}) = \sum_{i=1}^{n} \int_{\mathbf{I}} g_{jk}^{(i)}(\mathbf{v},\mathbf{w};\mathbf{u})\, d\mathbf{u}\,. \tag{10.3.13}$$

Summing with respect to the index i, Eq. (10.3.9) formally yields the total number of individuals evolution equation

$$\frac{dN}{dt} = \sum_{j,k=1}^{n} \int_{\mathbf{I}\times\mathbf{I}} (G_{jk}(\mathbf{v},\mathbf{w}) - 1)\, \eta_{jk}(\mathbf{v},\mathbf{w})\, f_j(t,\mathbf{v})\, f_k(t,\mathbf{w})\, d\mathbf{v}\, d\mathbf{w}\,. \tag{10.3.14}$$

The following (non exhaustive) particular cases can be assessed with reference to the above general class of models:

- ***Globally conservative models***:

$$G_{jk}(\mathbf{v},\mathbf{w}) = 1\,, \qquad \forall\, j,k = 1,\ldots,n\,, \qquad \forall\, \mathbf{v},\mathbf{w} \in \mathbf{I}\,.$$

  In this case the total number of individuals $N(t)$ can (formally) only remain equal to the number of individuals N_0 at $t = 0$.

- ***Globally proliferative models***:

$$G_{jk}(\mathbf{v},\mathbf{w}) > 1\,, \qquad \forall\, j,k = 1,\ldots,n\,, \qquad \forall\, \mathbf{v},\mathbf{w} \in \mathbf{I}\,.$$

  In this case the total number of individuals $N(t)$ can (formally) only increase

$$t \uparrow \quad \Rightarrow \quad N(t) \uparrow\,.$$

- ***Globally destructive models***:

$$G_{jk}(\mathbf{v},\mathbf{w}) < 1\,, \qquad \forall\, j,k = 1,\ldots,n\,, \qquad \forall\, \mathbf{v},\mathbf{w} \in \mathbf{I}\,.$$

In this case the total number of individuals $N(t)$ can (formally) only decrease

$$t \uparrow \quad \Rightarrow \quad N(t) \downarrow .$$

Qualitative analysis can be developed even in this general case. More detailed results can then be obtained for specific models. Consider the initial-value problem

$$\begin{cases} \dfrac{\partial \mathbf{f}}{\partial t} = J[\mathbf{f}] , \\ \mathbf{f}(0,\cdot) = \mathbf{f}_0(\cdot) , \end{cases} \tag{10.3.15}$$

for the functions

$$\mathbf{f} : t \in \mathbb{R}_+ \mapsto \mathbf{f}(t,\cdot) \in \mathbf{X} , \qquad \text{where} \quad \mathbf{X} := \big(L_1(\mathbf{I})\big)^n . \tag{10.3.16}$$

$\mathbf{X}$ is a Banach space when endowed with the norm

$$\|\mathbf{h}\| = \sum_{i=1}^{n} \int_{\mathbf{I}} |h_i(\mathbf{u})| \, d\mathbf{u} \, ; \qquad h_i \in L_1(\mathbf{I}) . \tag{10.3.17}$$

The analysis can be developed under the following

Assumption 10.3.6. *There exist $c_1, c_2 \in \mathbb{R}_+$ such that η_{ij} and $g_{jk}^{(i)}$, for $i, j, k = 1, \ldots, n$, satisfy*

$$0 \le \eta_{ij}(\mathbf{u}, \mathbf{v}) \le c_1 , \tag{10.3.18}$$

and

$$0 \le g_{jk}^{(i)}(\mathbf{v}, \mathbf{w}; \mathbf{u}) \le c_2 , \tag{10.3.19}$$

for Lebesgue-almost all $\mathbf{u}, \mathbf{v}, \mathbf{w} \in \mathbf{I}$.

It is easy to see that the operator $\mathbf{J} := \{J_i\}_{i=1}^n$ is locally Lipschitz continuous in $\mathbf{X}$ (the Lipschitz constant depends on c_1 and c_2) and hence there exists a unique solution $\mathbf{f}$ in $\mathbf{X}$ of the Problem (10.3.15) on a time interval $[0,T]$, where $T > 0$ only depends on c_1, c_2 and $\|\mathbf{f}_0\|$. Moreover, both operators $\mathbf{G} := \{G_i\}_{i=1}^n$ and $\mathbf{L} := \{L\}_{i=1}^n$ are positive and monotone. Therefore, by standard arguments already used in the theory of the Boltzmann equation, one can see that the solution $\mathbf{f}$ is nonnegative, provided that the initial datum is non-negative (we adhere to the convention that $\mathbf{f} \geq 0$ if $f_i \geq 0$ for every $i = 1,\ldots,n$). Local existence follows:

THEOREM 10.3.1. *Let Assumption 10.3.6 be satisfied. Then, for every $\mathbf{f}_0 \geq 0$ in $\mathbf{X}$, there exists $T > 0$ and a unique, non-negative, strong solution $\mathbf{f}(t)$ of Problem (10.3.15) in $\mathbf{X}$, for $t \in [0,T]$.*

In the general case, the solution cannot exist globally in time, due to possible divergences ([ARa]). However, in the globally conservative and globally destructive cases, we have the following a priori estimate

$$\|\mathbf{f}(t,\cdot)\| \leq \|\mathbf{f}_0\|\,, \qquad \text{for} \quad t \geq 0\,, \tag{10.3.20}$$

which guarantees global existence:

THEOREM 10.3.2. *Let Assumption 10.3.6 be satisfied and let the functions G_{jk}, defined by Eq. (10.3.13), be such that*

$$G_{jk}(\mathbf{v},\mathbf{w}) \leq 1\,, \qquad \forall j,k = 1,\ldots,n\,, \quad \text{and a.a. } \mathbf{v},\mathbf{w} \in \mathbf{I}\,. \tag{10.3.21}$$

Then there exists a unique, non-negative, solution $\mathbf{f} : [0,\infty) \to \mathbf{X}$ of Problem (10.3.15) for every $\mathbf{f}_0 \geq 0$ in $\mathbf{X}$. Moreover, inequality (10.3.20) is satisfied.

Some technical generalizations can be developed. For instance one can deal with ***models with internal structure***, i.e., such that the state variable follows an evolution equation determined by internal and/or external actions:

$$\frac{\partial f_i}{\partial t} + \sum_{j=1}^{n} \frac{\partial}{\partial u_j}\left(k_j(t,\mathbf{u})f_i\right) = J_i[\mathbf{f}]\,, \qquad k_j = \frac{du_j}{dt}\,, \tag{10.3.22}$$

where $k_i = k_j(t, \mathbf{u})$ is the activation term.

Practical situations may happen such that the interactions between individuals generate their outcome only after a certain delay time, to be modelled by appropriate memory terms. Then one has ***models with time structure.*** They can be designed by means of delay equations based on the assumption that changes of state, or of population, due to binary interactions, occur after suitable ***delay times***

$$\Delta_{ijk} \in \mathbf{R}_+ , \qquad i, j, k \in \{1, \ldots, n\} . \tag{10.3.23}$$

In this case the evolution equation reads

$$\begin{aligned} &\frac{\partial}{\partial t} f_i(t, \mathbf{u}) + \sum_{j=1}^{n} \frac{\partial}{\partial u_j} \left(k_j(t, \mathbf{u}) f_i(t, \mathbf{u}) \right) \\ &= \sum_{j,k=1}^{n} \int_{\mathbf{I} \times \mathbf{I}} \eta_{jk}(\mathbf{v}, \mathbf{w}) \, g_{jk}^{(i)}(\mathbf{v}, \mathbf{w}; \mathbf{u}) \\ &\times \; f_j(t - \Delta_{ijk}, \mathbf{v}) \, f_k(t - \Delta_{ijk}, \mathbf{w}) \, d\mathbf{v} \, d\mathbf{w} \\ &- \sum_{j=1}^{n} f_i(t - \Delta'_{ij}, \mathbf{u}) \int_{\mathbf{I}} \eta_{ij}(\mathbf{u}, \mathbf{v}) \, f_j(t - \Delta'_{ij}, \mathbf{v}) \, d\mathbf{v} , \end{aligned} \tag{10.3.24}$$

for $i = 1, \ldots, n$, and where $f_i : t \mapsto f_i(t, \cdot)$ is a given function on $[-\Delta, 0]$, where $\Delta = \max_{i,j,k} \{\Delta_{ijk}, \Delta'_{ij}\}$.

Dealing with ***models with space structure*** means that their state variable $\mathbf{u}$ includes the space variable. Hence, in addition, one has to model how interactions modify the interacting individuals velocities. The map between pre-interaction and post-interaction velocities may not necessarily obey the classical conservation equations of mechanics. Nevertheless, a conceivable modelling may only lead to a system evolution developed along lines analogous to those of the Boltzmann equation.

The reader may realize that the various models proposed in the preceding chapters may take a proper place in the above outlined framework. This is to say that all the models presented along the book can be regarded as particular cases of the general picture we discussed here above. Moreover, suitable developments of the various models may be studied by means of the generalizations of the global last one. On the other hand, this should not be seen as a simple technical exercise. A deep analysis is needed to adapt the general picture to the various specific cases, and it has to be regarded as a research perspective.

10.4 Development of New Models

Kinetic models can be developed whenever the real system to be simulated is constituted by a large number of indistinguishable interacting objects or subjects. The design of the mathematical models is developed, as we have seen, starting from a detailed analysis and simulation of the (microscopic) interactions between individuals, followed by the derivation of an evolution equation for the statistical distribution with respect to the state variable that describes the physical state of the system elements.

A general structure, and the modelling procedure, has been discussed in Section 10.2. However, the analysis of the interactions between individuals may substantially differ from system to system. In particular it can be noticed that if the elements of the large system are inert objects, then interactions can be modelled on the basis of mechanistic, hopefully deterministic laws. On the other hand, if the elements possess features that may be related to free will or thinking activities, such as those of the models dealt with in Chapter 3 on population dynamics or Chapter 8 on traffic flow, interactions have to be related not only to mechanics, but also to personal (even psychological) behaviors. In this case the determinism is lost, and stochasticity has to be taken into account.

Concerning now the modelling aspects, it is not too difficult to present a listing of further conceivable application fields. For instance one can suggest to take into account models with quantum interactions, e.g., [KEa] and [MAa], which have not been considered here, or models of granular media, communication networks etc. Some of these topics are dealt with in the surveys delivered in [BEc]. However, rather than identifying new fields of applications, we prefer to give research perspectives based on the idea that modelling in a certain environment may take substantial advantage from the knowledge in parallel environments. Hopefully, this will lead to a relatively richer model.

Along these lines the following indications, among several conceivable others, are suggested to the reader's attention:

- Condensation-fragmentation models, dealt with in Chapter 4, may contribute to the development of models of cell aggregation and condensation in the framework of cellular kinetic models dealt with in Chapter 5.
- Models of gas mixtures with inelastic collisions may contribute to the development of condensation fragmentation models in space dependent phenomena, which were dealt with in Chapter 4 only in the spatially homogeneous case.
- Mathematical models related to the generalized shape, dealt with in Chapter 6, may contribute to develop models of cellular interactions needed by the models dealt with in Chapter 5.
- Traffic flow models, dealt with in Chapter 8, may take advantage both of the Boltzmann equation generalizations to mixtures or inelastic collisions and of the population dynamics models dealt with in Chapter 3.

The above indications have to be regarded simply as examples, no completeness is claimed. Nevertheless, we claim that a deep insight into the correlation among the various kinetic models we are aware of is not yet available in the literature.

10.5 Closure

This book has been devoted to develop the methodological aspects and specific analysis of the various mathematical problems that arise when modelling large systems of identical elements with the methods of phenomenologic kinetic theory. Modelling and applications have been motivated by several interesting problems, and have generated interesting problems both of analytic and computational nature. Open problems have also been outlined, and showed to the attention of applied mathematicians.

The guideline, followed in the book, is that this new methodological approach can be useful when the traditional approach, which is typical of the phenomenologic mathematical physics, eventually fails. The growing attention raised in this way of modelling large systems is documented not only in the review paper [BEb], but also in the forthcoming collection of surveys [BEc] delivered by applied mathematicians who contributed to the developing of several kinetic models in applied sciences.

We suggest to consider this approach as a new modelling method, possibly able to generate a new class of equations in the framework of mathematical physics, that describes large systems of identical elements with the same mathematical structures that are of use in nonequilibrium statistical mechanics.

Within the above framework, one should also examine the links between the above statistical (microscopic) description and the traditional macroscopic modelling. This topic is well established for the classical Boltzmann equation. Following the procedure reviewed in Chapter 7, suitable limiting procedure (or, in physical terms, when the Knudsen number tends to zero) generates the macroscopic hydrodynamic equations. In general, suitable limiting procedures should lead to the corresponding macroscopic models. This problem, however difficult, is a challenging research perspective. Dealing with it will certainly contribute to a deeper understanding of the related kinetic and macroscopic models.

10.6 References

[ADa] ADAM J.A. and BELLOMO N. Eds., **A Survey of Models on Tumor Immune Systems Dynamics**, Birkhäuser, Basel, (1996).

[ARa] ARLOTTI L. and LACHOWICZ M., Qualitative analysis of a nonlinear integrodifferential equation modelling tumor-host dynamics, *Mathl. Comp. Modelling - Special Issue on Modelling and Simulation Problems on Tumor–Immune System Dynamics*, Bellomo N. Ed., **23** (1996), 11–30.

[ARb] ARLOTTI L., BELLOMO N., and LACHOWICZ M., Kinetic equations modelling population dynamics, *Transp. Theory Stat. Phys.*, to appear.

[ARc] ARLOTTI L., BELLOMO N., and LATRACH K., From the Jager and Segel model to kinetic population dynamics: Nonlinear evolution problems and applications, *Mathl. Comp. Modelling*, (1999), to appear.

[BEa] BELLOMO N. and GUSTAFSSON T., The discrete Boltzmann equation: A review of the mathematical aspects of the initial and initial-boundary value problem, *Review Math. Phys.*, **3** (1992), 137–162.

[BEb] BELLOMO N. and LO SCHIAVO M., From the Boltzmann equation to generalized kinetic models in applied sciences, *Mathl. Comp. Modelling*, **26**, (1997), 43–76.

[BEc] BELLOMO N. and PULVIRENTI M. Eds., **Modelling in Applied Sciences: A Kinetic Theory Approach**, Birkhäuser, Basel, (1999).

[BEd] BELLOMO N., FIRMANI B., and GUERRI L., Bifurcation analysis for a nonlinear system of integrodifferential equations modelling tumor immune cells competition, *Appl. Math. Letters*, **12**, (1999), 39–44.

[FIa] FIRMANI B., GUERRI L., and PREZIOSI L., Tumor immune system competition with medically induced activation disactivation, *Math. Models Meth. Appl. Sci.*, **9**, (1999), 491–512.

[GAa] GATIGNOL R., **Theorie Cinetique des Gaz a Repartition Discrete de Vitesses**, Lect. Notes in Phys. n.36, Springer, Heidelberg, (1975).

[GAb] GATIGNOL R., **Kinetic theory boundary conditions for discrete velocity gases**, *Phys. Fluids.*, **20** (1977), 2022–2030.

[KEa] KERSCH A. and MOROKIFF J., **Transport Simulation in Microelectronics**, Birkhäuser, Basel, (1989).

[LOa] LO SCHIAVO M., Discrete kinetic cellular models of tumor immune system interactions, *Math. Models Meth. Appl. Sci.*, **6**, (1996), 1187–1210.

[MAa] MARKOWICH P., RINGHOFER C., and SCHMEISER C., **Semiconductor Equations**, Springer, Heidelberg, (1990).

[MOa] MONACO R. and PREZIOSI L., **Fluid Dynamic Applications of the Discrete Boltzmann Equation**, World Scientific, London, Singapore, (1991).

[PLa] PLATKOWSKI T. and ILLNER R., Discrete velocity models of the Boltzmann equation: A survey on the mathematical aspects of the theory, *SIAM Review*, **30** (1988), 213–255.

Books and Review Papers

[5] ABBAS A.K., LICHTMANN A.H., and POBER J.S., **Cellular and Molecular Immunology**, Saunders, (1991).

[1,5] ADAM J.A. and BELLOMO N. eds., **A Survey of Models on Tumor Immune Systems Dynamics**, Birkhäuser, Boston, (1996).

[2] ANTOSIK P., MIKUSINSKI J., and SIKORSKI R., **Theory of Distributions**, Elsevier, (1973).

[5,7] ARLOTTI L. and BELLOMO N., On the Cauchy problem for the nonlinear Boltzmann equation, in **Lecture Notes on the Mathematical Theory of the Boltzmann Equation**, Bellomo N. ed., World Scientific, London, Singapore, (1995), 1–64.

[10] ARLOTTI L., BELLOMO N., and LATRACH K., From the Jager and Segel model to kinetic population dynamics: Nonlinear evolution problems and applications, *Mathl. Comp. Modelling*, **30**, (1999), 15–40.

[5] ASACHENKOV A., MARCHUK G., MOHOLER R., and ZUEV S., **Disease Dynamics**, Birkhäuser, Basel, (1994).

[2,4] ASH R. B., **Real Analysis and Probability**, Academic Press, New York, (1970).

[3] BELLENI-MORANTE A., **Applied Semigroups and Evolution Equations**, Clarendon Press, New York, (1980).

[3,8] BELLENI-MORANTE A. and MCBRIDE A., **Applied Nonlinear Semigroups**, Wiley, New York, (1998).

[7] BELLOMO N., PALCZEWSKI A., and TOSCANI G., **Mathematical Topics in Nonlinear Kinetic Theory**, World Scientific, London, Singapore, (1988).

[1,7,8] BELLOMO N., LACHOWICZ M., POLEWCZAK J., and TOSCANI G., **Mathematical Topics in Nonlinear Kinetic Theory II: The Enskog Equation**, World Scientific, London, Singapore, (1991).

[10] BELLOMO N. and GUSTAFSSON T., The discrete Boltzmann equation: A review of the mathematical aspects of the initial and initial-boundary value problem, *Review Math. Phys.*, **3** (1992), 137–162.

[1,7,9] BELLOMO N. ed., **Lecture Notes on the Mathematical Theory of the Boltzmann Equation**, World Scientific, London, Singapore, (1995).

[7,8] BELLOMO N., LE TALLEC P., and PERTHAME B., On the Solution of the Nonlinear Boltzmann Equation, *ASME Review*, **48** (1995), 777–794.

[3,8] BELLOMO N. and PREZIOSI L., **Modelling Mathematical Methods and Scientific Computation**, CRC Press, Boca Raton (FL), (1995).

[1,3,10] BELLOMO N. and LO SCHIAVO M., From the Boltzmann equation to generalized kinetic models in applied sciences, *Mathl. Comp. Modelling*, **26** (1997), 43–76.

[5] BELLOMO N. and DE ANGELIS, Strategies of applied mathematics towards an immuno-mathematical theory on tumors and immune system interactions, *Math. Models Meth. Appl. Sci.*, **8**, (1998), 1403–1429.

[1,10] BELLOMO N. and PULVIRENTI M. eds., **Modeling in Applied Sciences: A Kinetic Theory Approach**, Birkhäuser, Boston, (1999).

[7] BIRD G.A., **Molecular Gas Dynamics**, Oxford University Press, Oxford, (1976).

[7] BIRD G.A., **Molecular Gas Dynamics and the Direct Simulation of Gas Flows**, Clarendon Press, New York, (1994).

[1,7] CERCIGNANI C., ILLNER R. and PULVIRENTI M., **Theory and Application of the Boltzmann Equation**, Springer, Heidelberg, (1993).

[7] CERCIGNANI C., **Theory and Application of the Boltzmann Equation**, Springer, Heidelberg, (1988).

[7,8] CHORIN A. and MARSDEN J., **A Mathematical Introduction to Fluid Dynamics**, Springer, Heidelberg, (1979).

[5] CURTI B.D. and LONGO D.L., A brief history of immunologic thinking: It is time for Yin and Yang, in **A Survey of Models on Tumor Immune Systems Dynamics**, Adam J. and Bellomo N. eds., Birkhäuser, Boston, (1996), 1–14.

[5] DEN OTTER W. and RUITENBERG E.J. eds., **Tumor Immunology. Mechanisms, Diagnosis, Therapy**, Elsevier, New York, (1987).

[2] DUNFORD N. and SCHWARTZ J. T. **Linear Operators**, J. Wiley, London, (1988).

[4] EDWARDS R.E., **Functional Analysis: Theory and Applications**, Holt, Rinehart and Winston, New York, (1965).

[10] GATIGNOL R., **Theorie Cinetique des Gaz a Repartition Discrete de Vitesses**, Lect. Notes in Phys. n.36, Springer, Heidelberg, (1975).

[1,7] GLASSEY R., **The Cauchy Problem in Kinetic Theory**, SIAM Publ., Philadelphia, (1995).

[5] GREEN I., COHEN S., and MCCLUSKEY R. eds., **Mechanisms of Tumor Immunity**, Wiley, London, New York, (1977).

[7] GREENBERG W., ZWEIFEL P., and POLEWCZAK J., Global existence proofs for the Boltzmann equation, in **Nonlinear**

Phenomena: The Boltzmann Equation, Lebowitz J. and Montroll E. eds., North-Holland, Amsterdam, (1983), 21–49.

[7] GREENBERG W., VAN DER MEE C.V.M., and PROTOPOPESCU V., **Boundary Value Problems in Abstract Kinetic Theory**, Birkhäuser, Basel, (1987).

[9] GRÜNFELD C., Nonlinear kinetic models with Chemical reactions, in **Modeling in Applied Sciences, a Kinetic Theory Approach**, Bellomo N. and Pulvirenti M. eds., Birkhäuser, Boston, (1999).

[2] HALOMS P.R., **Measure Theory**, Van Nostrand, Princeton N.J., (1965).

[3] HOPPENSTEAD F., Mathematical Theory of Populations: Demographic, Genetics and Epidemics, *SIAM Conf. Series*, **2**, (1975).

[9] JABIN P. and PERTHAME B., Notes mathematical problems on the dynamics of dispersed particles, in **Modeling in Applied Sciences: A Kinetic Theory Approach**, Bellomo N. and Pulvirenti M. eds., Birkhäuser, Boston, (1999).

[7] KAPER H., LEKKERKERKER C. and HEJTMANEK J., **Spectral Theory in Linear Transport Equation**, Birkhäuser, Basel, (1982).

[8] KATO T., Nonlinear evolution equations in Banach spaces, **Proceedings of AMS Symposium in Applied Mathematics**, AMS, **17**, (1965), 50–67.

[2] KELLEY L., **General Topology**, GTM 27, Springer, Heidelberg, (1975).

[10] KERSCH A. and MOROKIFF J., **Transport Simulation in Microelectronics**, Birkhäuser, Basel, (1989).

[8] KLAR A., KÜNE R. D., and WEGENER R., Mathematical models for vehicular traffic, *Surveys Math. Ind.*, **6**, (1996) 215–239.

[8] KLAR A. and WEGENER R., Kinetic traffic flow models, in **Modeling in Applied Sciences: A kinetic Theory Approach**, Bellomo N. and Pulvirenti M. eds., (1999).

[2] KOGAN M.N., **Rarefied gas dynamics**, Plenum Press, New York, (1969).

[2] KOLOMGOROV A. N., and FOMIN S. V., **Introductory Real Analysis**, Dover, New York, (1975).

[1,7,8] LACHOWICZ M., Asymptotic analysis of nonlinear kinetic equations: The hydrodynamic limit, in **Lecture Notes on the Mathematical Theory of the Boltzmann Equation** , Bellomo N. ed., World Scientific, London, Singapore, (1995), 65—148.

[6] LAKSHMIKANTHAN V. and LEELA S., **Differential and Integral Inequalities**, Academic Press, New York, (1969).

[8] LEUTZBACK W., **Introduction to the Theory of Traffic Flow**, Springer, Heidelberg, (1988).

[2] LOEVE M., **Probability Theory**, GTM 46, Springer, Heidelberg, (1978).

[3] MACDONALD N., **Biological Delay Systems**, Cambridge Univ. Press, Cambridge, (1992).

[1,10] MARKOWICH A., RINGHOFER C., and SCHMEISER C., **Semiconductor Equations**, Springer, Heidelberg, (1990).

[1,7] MASLOVA N., **Nonlinear Evolution Equations**, World Scientific, London, Singapore, (1993).

[10] MONACO R. and PREZIOSI L., **Fluid Dynamic Applications of the Discrete Boltzmann Equation**, World Scientific, London, Singapore, (1991).

[5] MURRAY J., **Mathematical Biology**, Springer, Heidelberg, (1994).

[1,7] NEUNZERT H. and STRUCKMEIER J., Particle Methods for the Boltzmann equation, in **Acta Numerica 1995**, Cambridge University Press, (1995), 417-458.

[5] NOSSAL G.J., Life, death and the immune system, *Scientific American*, **269** (1993), 53–72.

[10] PAZY A., **Semigroups of Linear Operators and Applications to Partial Differential Equations**, Springer, Heidelberg, (1982).

[10] PLATKOWSKI T. and ILLNER R., Discrete velocity models of the Boltzmann equation: A survey on the mathematical aspects of the theory, *SIAM Review*, **30** (1988), 213–255.

[1,8] PRIGOGINE I. and HERMAN R., **Kinetic Theory of Vehicular Traffic**, Elsevier, New York, (1971).

[2] ROYDEN H. L., **Real Analysis**, Mac Millan, New York, (1968).

[2] RUELLE D., **Statistical Mechanics, Rigorous Results**, W.A. Benjamin, Reading Mass., (1969).

[1,5] SEGEL L., **Modelling Dynamic Phenomena in Molecular and Cellular Biology**, Cambridge University Press, Cambridge, (1984).

[3] STREATER R.F., **Statistical Dynamics**, Imperial College Press, (1995).

[1,7] TRUESDELL C. and MUNCASTER R., **Fundamentals of Maxwell Kinetic Theory of a Simple Monoatomic Gas**, Academic Press, New York, (1980).

[7] WALUS W., Current computational methods for the nonlinear Boltzmann equation, in **Lecture Notes on the Mathematical Theory of the Boltzmann Equation**, Bellomo N. ed., World Scientific, London, Singapore, (1995).

[8] WHITHAM G., **Linear and Nonlinear Waves**, J. Wiley, London (1978).

[6] ZEIDER E., **Applied Functional Analysis**, Springer, Heidelberg, 1995.

Series on Advances in Mathematics for Applied Sciences

Aims and Scope

This Series reports on new developments in mathematical research relating to methods, qualitative and numerical analysis, mathematical modeling in the applied and the technological sciences. Contributions related to constitutive theories, fluid dynamics, kinetic and transport theories, solid mechanics, system theory and mathematical methods for the applications are welcomed.

This Series includes books, lecture notes, proceedings, collections of research papers. Monograph collections on specialized topics of current interest are particularly encouraged. Both the proceedings and monograph collections will generally be edited by a Guest editor.

High quality, novelty of the content and potential for the applications to modern problems in applied science will be the guidelines for the selection of the content of this series.

Instructions for Authors

Submission of proposals should be addressed to the editors-in-charge or to any member of the editorial board. In the latter, the authors should also notify the proposal to one of the editors-in-charge. Acceptance of books and lecture notes will generally be based on the description of the general content and scope of the book or lecture notes as well as on sample of the parts judged to be more significantly by the authors.

Acceptance of proceedings will be based on relevance of the topics and of the lecturers contributing to the volume.

Acceptance of monograph collections will be based on relevance of the subject and of the authors contributing to the volume.

Authors are urged, in order to avoid re-typing, not to begin the final preparation of the text until they received the publisher's guidelines. They will receive from World Scientific the instructions for preparing camera-ready manuscript.

SERIES ON ADVANCES IN MATHEMATICS FOR APPLIED SCIENCES

Vol. 17 The Fokker-Planck Equation for Stochastic Dynamical Systems and Its Explicit Steady State Solution
by C. Soize

Vol. 18 Calculus of Variation, Homogenization and Continuum Mechanics
eds. G. Bouchitté et al.

Vol. 19 A Concise Guide to Semigroups and Evolution Equations
by A. Belleni-Morante

Vol. 20 Global Controllability and Stabilization of Nonlinear Systems
by S. Nikitin

Vol. 21 High Accuracy Non-Centered Compact Difference Schemes for Fluid Dynamics Applications
by A. I. Tolstykh

Vol. 22 Advances in Kinetic Theory and Computing: Selected Papers
ed. B. Perthame

Vol. 23 Waves and Stability in Continuous Media
eds. S. Rionero and T. Ruggeri

Vol. 24 Impulsive Differential Equations with a Small Parameter
by D. Bainov and V. Covachev

Vol. 25 Mathematical Models and Methods of Localized Interaction Theory
by A. I. Bunimovich and A. V. Dubinskii

Vol. 26 Recent Advances in Elasticity, Viscoelasticity and Inelasticity
ed. K. R. Rajagopal

Vol. 27 Nonstandard Methods for Stochastic Fluid Mechanics
by M. Capinski and N. J. Cutland

Vol. 28 Impulsive Differential Equations: Asymptotic Properties of the Solutions
by D. Bainov and P. Simeonov

Vol. 29 The Method of Maximum Entropy
by H. Gzyl

Vol. 30 Lectures on Probability and Second Order Random Fields
by D. B. Hernández

Vol. 31 Parallel and Distributed Signal and Image Integration Problems
eds. R. N. Madan et al.

Vol. 32 On the Way to Understanding The Time Phenomenon: The Constructions of Time in Natural Science – Part 1. Interdisciplinary Time Studies
ed. A. P. Levich

SERIES ON ADVANCES IN MATHEMATICS FOR APPLIED SCIENCES

Vol. 33 Lecture Notes on the Mathematical Theory of the Boltzmann Equation
ed. N. Bellomo

Vol. 34 Singularly Perturbed Evolution Equations with Applications to Kinetic Theory
by J. R. Mika and J. Banasiak

Vol. 35 Mechanics of Mixtures
by K. R. Rajagopal and L. Tao

Vol. 36 Dynamical Mechanical Systems Under Random Impulses
by R. Iwankiewicz

Vol. 37 Oscillations in Planar Dynamic Systems
by R. E. Mickens

Vol. 38 Mathematical Problems in Elasticity
ed. R. Russo

Vol. 39 On the Way to Understanding the Time Phenomenon: The Constructions of Time in Natural Science — Part 2. The "Active" Properties of Time According to N. A. Kozyrev
ed. A. P. Levich

Vol. 40 Advanced Mathematical Tools in Metrology II
eds. P. Ciarlini et al.

Vol. 41 Mathematical Methods in Electromagnetism Linear Theory and Applications
by M. Cessenat

Vol. 42 Constrained Control Problems of Discrete Processes
by V. N. Phat

Vol. 43 Motor Vehicle Dynamics: Modeling and Simulation
by G. Genta

Vol. 44 Microscopic Theory of Condensation in Gases and Plasma
by A. L. Itkin and E. G. Kolesnichenko

Vol. 45 Advanced Mathematical Tools in Metrology III
eds. P. Cianlini et al.

Vol. 46 Mathematical Topics in Neutron Transport Theory — New Aspects
by M. Mokhtar-Kharroubi

Vol. 47 Theory of the Navier-Stokes Equations
eds. J. G. Heywood et al.

Vol. 48 Advances in Nonlinear Partial Differential Equations and Stochastics
eds. S. Kawashima and T. Yanagisawa

SERIES ON ADVANCES IN MATHEMATICS FOR APPLIED SCIENCES

Vol. 49 Propagation and Reflection of Shock Waves
by F. V. Shugaev and L. S. Shtemenko

Vol. 50 Homogenization
eds. V. Berdichevsky, V. Jikov and G. Papanicolaou

Vol. 51 Lecture Notes on the Mathematical Theory of Generalized Boltzmann Models
by N. Bellomo and M. Lo Schiavo

www.ingramcontent.com/pod-product-compliance
Ingram Content Group UK Ltd.
Pitfield, Milton Keynes, MK11 3LW, UK
UKHW022108170926
8624IPUK00024B/256

9 789810 240783